Building Code Basics: Residential

Based on the 2009 International Residential Code®

Building Code Basics: Residential

Based on the 2009 International Residential Code®

International Code Council
Stephen A. Van Note, CBO

DELMAR
CENGAGE Learning™

Australia • Brazil • Japan • Korea • Mexico • Singapore • Spain • United Kingdom • United States

**Building Code Basics: Residential,
Based on the 2009 International
Residential Code®**
Stephen A. Van Note

Vice President, Technology and Trades
   Professional Business Unit:
   Gregory L. Clayton

Product Development Manager: Ed Francis

Development: Dawn M. Jacobson

Editorial Assistant: Nobina Chakraborti

Director of Marketing: Beth A. Lutz

Executive Marketing Manager:
   Taryn Zlatin McKenzie

Marketing Manager: Marissa Maiella

Production Director: Carolyn Miller

Production Manager: Andrew Crouth

Senior Content Project Manager:
   Kara A. DiCaterino

Art Director: Benjamin Gleeksman

ICC Staff:

Senior Vice President, Business and Product
   Development: Mark A. Johnson

Technical Director, Product Development:
   Doug Thornburg

Manager, Project and Special Sales:
   Suzane Nunes Holten

Senior Marketing Specialist:
   Dianna Hallmark

For product information and technology assistance, contact us at
**Professional Group Cengage Learning Customer &
Sales Support, 1-800-354-9706**

For permission to use material from this text or product,
submit all requests online at **cengage.com/permissions.**
Further permissions questions can be e-mailed to
permissionrequest@cengage.com.

Library of Congress Control Number: 2009933659

ISBN-13: 978-1-4354-0063-4

ISBN-10: 1-4354-0063-1

ICC World Headquarters
500 New Jersey Avenue, NW
6th Floor
Washington, D.C. 20001-2070
Telephone: 1-888-ICC-SAFE (422-7233)
Website: http://www.iccsafe.org

Delmar
5 Maxwell Drive
Clifton Park, NY 12065-2919
USA

Cengage Learning is a leading provider of customized learning solutions
with office locations around the globe, including Singapore, the United
Kingdom, Australia, Mexico, Brazil and Japan. Locate your local office at:
international.cengage.com/region

Cengage Learning products are represented in Canada by
Nelson Education, Ltd.

For more learning solutions, please visit our corporate website at
www.cengage.com

Visit us at **www.InformationDestination.com**

Notice to the Reader

Publisher does not warrant or guarantee any of the products described herein or perform any independent analysis in connection with
any of the product information contained herein. Publisher does not assume, and expressly disclaims, any obligation to obtain and
include information other than that provided to it by the manufacturer. The reader is expressly warned to consider and adopt all safety
precautions that might be indicated by the activities described herein and to avoid all potential hazards. By following the instructions
contained herein, the reader willingly assumes all risks in connection with such instructions. The publisher makes no representations or
warranties of any kind, including but not limited to, the warranties of fitness for particular purpose or merchantability, nor are any such
representations implied with respect to the material set forth herein, and the publisher takes no responsibility with respect to such
material. The publisher shall not be liable for any special, consequential, or exemplary damages resulting, in whole or part, from the
readers' use of, or reliance upon, this material.

Printed in the United States of America
1 2 3 4 5 XX 11 10 09

CONTENTS

PART I: CODE ADMINISTRATION AND ENFORCEMENT

PART IV: FINISHES AND WEATHER PROTECTION

PART V: HEALTH AND SAFETY

PART VI: BUILDING UTILITIES

PART VII: ENERGY CONSERVATION

PART VIII: PROTECTION FROM OTHER HAZARDS

PREFACE

Construction of residential buildings routinely consists of conventional practices, those tried-and-true methods that have performed well over the years and have long been recognized by the building code. With the introduction of new technology, materials, and methods, improved understanding of safe and healthy living environments, and innovation in dwelling designs, residential construction and the codes that regulate it have become increasingly complex. Such complexity is necessary to afford flexibility in design and construction. Reference publications intending to explain the provisions regulating residential construction may overwhelm the reader with a broad range of topics and alternatives, may provide superficial coverage of all requirements, or may focus on the details of a limited number of provisions.

Building Code Basics: Residential—Based on the 2009 International Residential Code® was specifically developed to address the need for an illustrated text explaining the basics of the residential code—those provisions essential to understanding the application of the code to the most commonly encountered building practices. The text is presented and organized in a user-friendly manner with an emphasis on technical accuracy and clear non-code language. The content is directed to readers with a basic understanding of conventional dwelling construction but a less than complete knowledge of the *International Residential Code®* (IRC).

Anyone involved in the design, construction, or inspection of residential buildings will benefit from this book. Beginning and experienced inspectors, contractors, home builders, architects, designers, home inspectors, and students of construction technology or related fields will gain a fundamental understanding and practical application of the frequently used provisions of the 2009 edition of the IRC.

The content of *Building Code Basics: Residential* is organized to correspond to the order of construction, beginning with sitework and foundations through completion of a safe, healthy, and energy-efficient dwelling. Mechanical, fuel-gas, plumbing, and electrical provisions are placed in separate chapters. The advantage of this format to the reader is that it pulls related information together from various sections of the IRC into one convenient location of the text and provides a familiar frame of reference to those with any construction experience. The book explains the difference between "prescriptive" and "performance" requirements. Prescriptive structural design requirements to resist the forces of wind, earthquake, and snow are described and illustrated in an easy to understand way. Structural topics include conventional footings and foundations (including the fundamentals of soil capacity), conventional wood floor, wall and roof framing, engineered wood products, and seismic reinforcing of masonry chimneys. Fire- and life-safety concerns are addressed with topics including means of egress, emergency escape, stairways, fall protection, smoke alarms, fire sprinklers, and fire resistant construction. *Building Code Basics: Residential* also covers the minimum interior environmental conditions for a healthy living environment, weather protection, and energy conservation measures.

Correct and reasonable application of the code provisions is enhanced by a basic understanding of the code development process, the scope, intent, and correlation of the family of International Codes, and the proper administration of those codes. Such fundamental information is provided in the opening chapters of this publication. The book also explains the interaction of a building code with other local and state regulations and includes discussion of common hazards of the built environment that may be regulated by state or federal agencies.

This book does not intend to cover all provisions of the IRC or all of the accepted materials and methods of construction of residential buildings. Focusing in some detail on the most common conventional construction provisions affords an opportunity to fully understand the basics without exploring every variable and alternative. This is not to say that information not covered is any less important or valid. This book is best used as a companion to the IRC, which should be referenced for more complete information.

Building Code Basics: Residential features full-color illustrations to assist the reader in visualizing the application of the code requirements. Practical examples, simplified tables, and highlights of particularly useful information also aid in understanding the provisions and determining code compliance. References to the applicable sections of the 2009 edition of the IRC are helpful in locating the corresponding code language and related topics in the code. A glossary of code and construction terms clarifies the meaning of the technical provisions.

ABOUT THE INTERNATIONAL RESIDENTIAL CODE

The IRC is a comprehensive, stand-alone residential code that establishes minimum regulations for the construction of one- and two-family dwellings and townhouses up to three stories in height, including provisions for fire and life safety, structural design, energy conservation, and mechanical, fuel-gas, plumbing, and electrical systems. The IRC incorporates prescriptive provisions for conventional construction as well as performance criteria that allow the use of new materials and new building designs.

The IRC is one of the codes in the family of *International Codes* published by the International Code Council (ICC). All are maintained and updated through an open code development process and are available internationally for adoption by the governing authority to provide consistent enforceable regulations for the built environment.

ACKNOWLEDGMENTS

Building Code Basics: Residential is the result of a collaborative effort, and the author is grateful for the valuable contributions by the following talented staff of ICC Product Development: Hamid Naderi, PE, Vice

President, developed the concept for the book, provided direction throughout the process, and provided an in-depth review of the manuscript. A special thank-you goes to John Henry, PE, ICC Principal Staff Engineer, for his generous assistance and patient explanations relating to the structural provisions. Beyond his well-known expertise, John is above all a teacher, and we as students benefit from his willingness to share. Scott Stookey, senior technical staff, provided welcome expertise on fire resistance and fire protection systems and related photographs. Thanks to Peter Kulczyk, senior technical staff, for his helpful comments and access to his photo library. Thanks also go to Doug Thornburg, AIA, Technical Director of Product Development, for his usual expert direction and advice. All contributed to the accuracy and quality of the finished product.

ABOUT THE AUTHOR

Stephen A. Van Note, CBO
International Code Council
Senior Technical Staff, Product Development

Stephen A. Van Note is a member of the senior technical staff of the International Code Council (ICC), where, as part of the Product Development team, he is responsible for authoring technical resource materials in support of the International Codes. His role also includes the management, review, and technical editing of publications authored by outside sources. Prior to joining ICC in 2009, Van Note was building official for Linn County, Iowa. He has 15 years of experience in code administration and enforcement, and over 20 years of experience in the construction field, including project planning and management for residential, commercial, and industrial buildings. A certified building official and plans examiner, Van Note also holds certifications in five inspection categories.

Code Administration and Enforcement

Introduction to Building Codes

Building codes, in the broadest sense, are the various sets of regulations related to the construction, alteration, maintenance, and use of buildings and structures. Sometimes collectively called construction codes, the separate volumes include not only structural considerations, but provisions for fire and life safety, energy conservation, and systems for heating, cooling, plumbing, and electrical utilities. These codes serve primarily to protect the safety and welfare of the building occupants and the public. One in a family of coordinated and compatible construction codes, the *International Residential Code* (IRC) combines all elements necessary for the construction of one- and two-family dwellings and townhouses into a single volume. Providing design flexibility, the IRC references companion International Codes for elements of construction outside the scope of the IRC. This chapter briefly discusses the code development process and the scope of some of the companion codes, followed by a more detailed examination of the IRC scope (Figure 1-1).

CODE DEVELOPMENT

Just as construction technology, methods, and materials are constantly changing, model codes too are constantly undergoing review and a process for updating at periodic intervals to keep up with the fast pace of change. The International Codes are revised and updated through an open process that invites participation by all stakeholders and often includes exhaustive research, review, discussion, and debate of the issues.

FIGURE 1-1 International Codes

A new edition of the code is published every three years and reflects the results of the code development hearings. A code change begins with the submittal of a proposal. Any interested individual or group may submit a code change proposal and participate in the proceedings in which it and all other proposals are considered. Following the publication and distribution of all proposals, an open hearing is held before a committee comprised of representatives from across the construction industry, including code regulators, contractors, builders, architects, engineers, and others with expertise related to the applicable code or portion of the code under consideration. This open debate and broad participation before the committee ensures a consensus of the construction community and those impacted by building codes in the decision-making process. The committee may approve, modify, or disapprove the code change proposal.

The ICC membership present at the hearing has the opportunity to overturn the vote of the committee. Following the published results of the hearing, anyone may submit a written public comment proposing to overturn or modify the hearing results. The next public hearing is the *final action hearing,* in which the merits of code change proposals that received public comment are debated. Though any interested party may offer testimony, only ICC government members (designated public safety officials of a government jurisdiction responsible for administering and enforcing the codes) are permitted to cast votes at the final action hearing. Public safety officials have no vested financial interest in the outcome and legitimately represent the public interest. This important process ensures that the International Codes will reflect the latest technical advances and address the concerns of those throughout the industry in a fair and equitable manner.

THE BUILDING CODES: SCOPE AND LIMITATIONS

There are a number of features common to all of the International Codes. Each code begins by stating its scope of application. The scope establishes the range of buildings, uses, construction, equipment, and systems to which the particular code applies. A purpose statement follows the scope and includes the intent to provide minimum standards to protect the health, safety, and welfare of the public and the occupant or user of

Code Basics

International Code development cycle:

1. Anyone can submit a code change proposal.
2. Proposals are printed and distributed.
3. Open public hearings are held before committee.
4. Public hearing results are printed and distributed.
5. Anyone can submit public comment on hearing results.
6. Public comments are printed and distributed.
7. An open public final action hearing is held.
8. Final votes are cast by ICC government members.
9. A new edition is published. ●

the space or building. Subsequent sections place limitations on the application of the code. For example, each code permits the continued legal use of existing buildings. A building does not need to be brought into compliance with the current codes provided the building does not create hazards to the occupants or property and meets minimum acceptable standards for health and sanitation. In the case of an addition to an existing building, for instance, only the addition need comply with the current code, provided it does not cause an unsafe condition in the existing structure. Each International Code also references other codes and standards for use under specific circumstances. For example, the IRC references the *International Building Code* (IBC) for an engineered design of structural elements beyond the scope of the IRC. Finally, the appendices of each code are not in effect unless they are specifically adopted by the local jurisdiction having authority (see Chapter 2).

International Building Code (IBC)

The provisions of the IBC apply to the construction, alteration, maintenance, use, and occupancy of all buildings and structures except detached one- and two-family dwellings and townhouses and their accessory structures, which are covered by the IRC. In addition to structural components and systems, the IBC provides for a safe means of egress, accessibility for persons with disabilities, fire resistance, fire protection systems, weather resistance, finishes, and interior environments. These regulations are typically related to the use and occupancy of the building. That is, the IBC assesses relative risks or hazards based on the functions within the building and controls design accordingly. As a result, provisions regulating the building's size, means of egress, fire resistive elements, and fire protection systems vary significantly among sports arenas, hospitals, schools, apartments, and office buildings (Figure 1-2).

International Mechanical Code (IMC)

The provisions of the *International Mechanical Code* (IMC) generally apply to the installation, alteration, use, and maintenance of permanent mechanical systems utilized for comfort heating, cooling, and ventilation (HVAC), and other mechanical processes within buildings. The IMC references the IRC for detached one- and two-family dwellings and townhouses and their accessory structures.

International Fuel Gas Code (IFGC)

The *International Fuel Gas Code* (IFGC) regulates the installation of natural gas and liquefied petroleum (LP)–gas piping systems, fuel gas appliances, gaseous hydrogen systems, and related accessories. The fuel gas piping system extends from the utility company's point of delivery to the appliance shutoff valves. Coverage includes pipe sizing and other design considerations, approved materials, installation, testing, inspection, operation, and maintenance. The equipment installation requirements include combustion

FIGURE 1-2 Apartment building constructed under the provisions of the IBC *(Courtesy of iStock)*

and ventilation air, approved venting, and connection to the fuel gas piping system. In addition to the code requirements, gas appliances must be listed and labeled as meeting the applicable standards, and installations are governed by the listing and the manufacturer's instructions. The IFGC excludes from its scope detached one- and two-family dwellings and townhouses and their accessory structures, and references the IRC for these buildings.

International Plumbing Code (IPC)

The provisions of the *International Plumbing Code* (IPC) generally apply to the installation, alteration, use, and maintenance of plumbing systems. The IPC includes the material and installation requirements for water supply and distribution, plumbing fixtures, drain-waste and vent (DWV) piping, and storm drainage systems. As with the IBC, IMC, and IFGC, the IPC references the IRC for regulation of detached one- and two-family dwellings and townhouses and their accessory structures.

International Fire Code (IFC)

The *International Fire Code* (IFC) contains requirements for protection of buildings, occupants, and fire responders, prevention of fires, and the accidental release of hazardous materials. Storage, building use, and building maintenance provisions apply during the life of buildings. The IFC also contains design and construction provisions related to fire department access, fire protection water supply, and fire alarm and fire protection systems. Building use regulations include hazardous processes, refrigeration systems, stationary lead acid battery systems, emergency power using diesel fuel, and LP gas–supplied emergency generators. The IFC establishes regulations for the prevention of fire and explosion hazards arising from the storage and use of hazardous materials and controls other conditions that are hazardous to life, property, or public welfare. Construction of detached one- and two-family dwellings and townhouses and their accessory structures are regulated by the IRC.

International Property Maintenance Code (IPMC)

The *International Property Maintenance Code* (IPMC) contains clear and specific property maintenance regulations. The purpose of the IPMC is to adequately protect the public safety, health, and general welfare as they are affected by the continued occupancy and maintenance of structures and premises. Existing structures and premises that do not comply with these provisions must be altered or repaired to provide a minimum level of health and safety. The provisions of the IPMC apply to all existing residential and nonresidential structures and all existing premises. They include minimum requirements for light, ventilation, space, heating, sanitation, protection from the elements, life safety, safety from fire and other hazards, and for safe and sanitary maintenance.

International Existing Building Code (IEBC)

The *International Existing Building Code* (IEBC) applies to the repair, alteration, change of occupancy, addition, and relocation of existing buildings. The intent is to provide flexibility and alternative approaches in the repair, alteration, and addition to existing buildings and to still safeguard the public health, safety, and welfare. The alternatives offered to the applicant include prescriptive, work-area, and performance-compliance methods. Unless the building has sustained substantial structural damage or is undergoing significant structural alteration, construction is permitted to comply with the laws in existence at the time the building was constructed. The IEBC references the IBC for the installation of new structural members in an existing building. The legal occupancy of an existing building is permitted to continue without change, except where the general safety and welfare of the occupants and the public is compromised.

International Energy Conservation Code (IECC)

The *International Energy Conservation Code* (IECC) regulates the design and construction of buildings for the effective use of energy. This code applies to residential and commercial buildings. The IECC intends to provide flexibility in the methods of compliance with energy conservation requirements. The use and occupancy of existing buildings is generally permitted to continue without upgrading to the IECC energy efficiency provisions. The IECC applies to additions and some alterations to existing buildings without requiring the entire building to be brought into compliance. Specific exemption from IECC requirements is granted to certified historic buildings.

INTERNATIONAL RESIDENTIAL CODE (IRC)

The provisions of the IRC generally apply to the construction, alteration, use, and occupancy of detached one- and two-family dwellings and townhouses (Figures 1-3 through 1-5). Such buildings are limited to not more than three stories above grade in height, and each dwelling unit must have a separate means of egress. This comprehensive, stand-alone residential code includes provisions for structural elements, fire and life safety, a healthy living environment, energy conservation, and mechanical, fuel gas, plumbing, and electrical systems. The IRC incorporates prescriptive provisions for conventional light frame construction as well as performance criteria that allow the use of new materials and new building designs (see Chapter 4). As in the other International Codes, the purpose of the IRC is to safeguard the public safety, health, and general welfare from fire and other potential hazards attributed to the built environment. The code establishes minimum requirements, reasonably balanced for affordability, that provide strong, stable, and sanitary homes that conserve energy while still offering adequate lighting, comfort conditioning, and ventilation. [Ref. R101, R202]

Dwellings and townhouses

The building height and means of egress requirements of the IRC apply equally to one- and two-family dwellings and townhouses. While the code generally limits these residential buildings to three stories above ground level, this still permits a full basement in addition to three stories above, effectively creating a building with four floor levels. In addition,

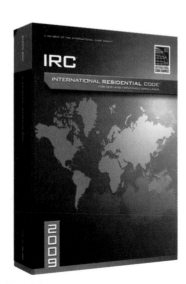

FIGURE 1-3 *2009 International Residential Code* (IRC)

Code Basics

Prescriptive code provisions:

- A set of rules to go by
- A recipe to follow
- Do this and do that according to a "prescription"

Examples of prescriptive provisions:

- Conventional wood framing span tables
- Braced wall panels for conventional wood framing
- Height of handrails and guards
- Stair dimensions

Performance code provisions:

- Systems or components must function in a certain way to meet the desired level of safety and performance
- Performance of structural systems is typically achieved through engineering
- Offers variable and multiple solutions for flexibility of design and construction
- Permits the use of alternate materials, equipment, and methods of construction

Examples of performance provisions:

- Design of engineered wood trusses
- Design of shear walls
- Sizing of steel beams and columns
- Graspability of handrails
- Strength of handrails, guards, and stairs ●

the 2009 IRC permits a habitable attic, which is not counted as a story, conceivably creating a fifth habitable level, though such an installation is not common (Figure 1-6). As will be seen in later chapters, structural and other design criteria of the code may further limit the height and number

FIGURE 1-4 Single-family dwelling

FIGURE 1-5 Townhouses *(Courtesy of iStock)*

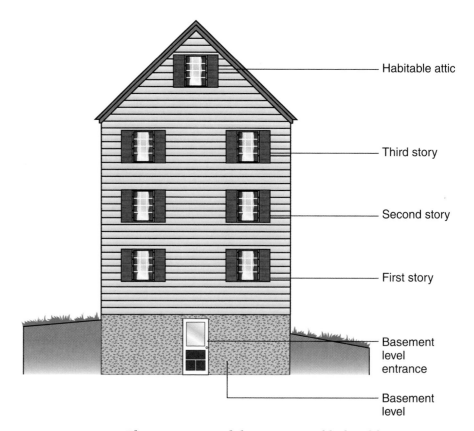

FIGURE 1-6 Three stories with basement and habitable attic

of stories of the building. The code does not limit the total area of dwellings, however.

In addition to height considerations, each dwelling unit requires its own separate means of exiting the building to the outdoors (see Chapter 8). Only one exterior exit door is required, and the travel distance to that exit is not regulated, no matter the size or number of stories of the dwelling unit. Two-family dwellings (Figure 1-7) and townhouses require fire resistant separations between dwelling units. Limited protection against the spread of fire is also required between a dwelling unit and an attached garage (see Chapter 9). [Ref. R101.2, R202]

The IRC does not limit the number of townhouses in a group of townhouses but does require the building to satisfy certain other conditions. To qualify as a townhouse, there must be at least three attached dwelling units, and each unit must run from foundation to roof. That is, any portion of a townhouse is not permitted to be placed above any portion of another townhouse. In addition, each townhouse must be open to a yard or public way on at least two sides (Figure 1-8). Multifamily dwellings that do not meet the definition of townhouses fall under the provisions of the IBC (Figure 1-9). [Ref. R202]

FIGURE 1-7 Two-family dwelling *(Courtesy of iStock)*

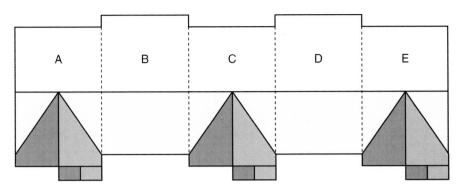

FIGURE 1-8 Townhouses open on front and back

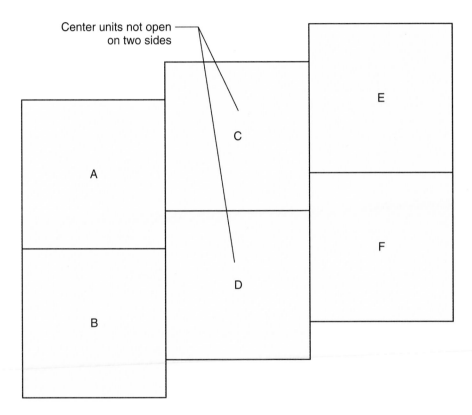

FIGURE 1-9 Six-unit multifamily dwelling outside the scope of the IRC

Manufactured homes

Regulations related to manufactured homes appear in Appendix E of the IRC and are not in effect unless specifically adopted by the jurisdiction. For purposes of the IRC, a manufactured home is considered the same as a mobile home, though the preferred term since 1974 federal legislation is *manufactured home*. The United States Department of Housing and Urban Development (HUD) regulates the construction of manufactured homes, which are built in a manufacturing plant to comply with the *Manufactured Home Construction and Safety Standards* (HUD

FIGURE 1-10 **Manufactured home** *(Courtesy of iStock)*

code). Compliance is verified through state or third party inspection agencies. Each transportable section must display a red certification label on the exterior.

Manufactured homes are built on a permanent chassis and are designed to be used as a dwelling with or without a permanent foundation (Figure 1-10). This design assures transportability for relocation and differentiates manufactured homes from modular homes and other factory-built, panelized, or component structures.

Local jurisdictions have no authority to regulate the design or construction of a HUD-regulated manufactured home. IRC Appendix E applies to on-site construction related to manufactured homes used as single-family dwellings and installed on privately owned lots. The provisions relate to the foundation system, connection to utility services (water, sewer, fuel, and electricity), and alterations, additions, or repairs to the manufactured home. However, any modification or addition must be in compliance with HUD regulations and cannot proceed if otherwise prohibited. Attachment of an accessory building, such as a garage, is generally prohibited unless the design is substantiated through engineering calculations. Factory-built structures that do not fall under the HUD rules are regulated by local jurisdictions like any other building, or in some cases are regulated by a state agency.

Accessory buildings and structures

Accessory buildings must be located on the same lot as the dwelling and are limited to two stories and not more than 3,000 square feet in floor area. In addition, the use of such buildings must be considered customarily accessory to and incidental to that of the dwelling. Detached garages and sheds are the most common examples of accessory buildings (Figure 1-11). Gazebos, playhouses, swimming pool equipment rooms, and garden buildings are other examples of structures typically considered accessory to dwellings. **[Ref. R202]**

Existing structures

Provisions allowing the legal occupancy of buildings to continue without fully complying with current codes are often referred to as grandfathering or grandfather clauses. As with other International Codes, the IRC provides such relief for existing buildings. To impose regulations to bring existing buildings into current compliance would be impractical and unreasonable and would penalize owners of buildings that complied with applicable laws at the time of their construction. Of course, if, due to lack of repair or maintenance, buildings fall below the generally acceptable threshold for sanitation, health, safety, and welfare of the occupants and the public, the IRC requires corrections in accordance with specific code provisions and the referenced provisions of the IPMC or the IFC.

These grandfathering provisions not only apply to the continued use and occupancy of an existing building in the absence of construction activity, they also apply to existing buildings undergoing modifications or additions. Generally, only the modification or addition need comply with the current code. There are some exceptions to this rule. For example, additional smoke alarms may be required as for new construction. (See Chapter 9.) The installation of smoke alarms, of paramount importance in saving lives, is considered reasonable and practical during construction work on existing buildings. Additions, alterations, or repairs cannot cause any portion of the existing building to become unsafe or otherwise adversely affect the performance of the building. If an addition impeded the means of egress to the outdoors, added excessive loading to existing structural members, overloaded the electrical service, or exceeded the capacity of the plumbing DWV system, then any of the affected elements would need to be brought into compliance with the current code.

IRC Appendix J, if specifically adopted, does offer alternatives for compliance with the code during renovation of existing buildings. These provisions, similar to those found in the IEBC, intend to encourage the continued use or reuse of legally existing buildings and structures. Under Appendix J, construction work is categorized as repair, renovation, alteration, or reconstruction. As the extent of work becomes greater, resulting in a higher category, the requirements become more stringent. While maintaining an acceptable level of safety, the alternative approaches offer a number of benefits to the owner and builder, such as allowing smaller dimensions for existing windows, doors, stair headroom, and ceiling heights without modification. In some cases, certain nonconforming elements in sound condition, such as stairs and railings, may remain without modification even within prescribed work areas. [Ref. R102.7]

FIGURE 1-11 Accessory building

Legal Aspects, Permits, and Inspections

Building codes intend to protect the health, safety, and welfare of the public by establishing minimum acceptable standards of construction. In order to be effective, such codes must be legally adopted by the governmental jurisdiction and enforced by qualified officials appointed by the governing authority. This chapter explores the process of adopting, amending, and administering the *International Residential Code* (IRC), but the principles generally apply to the other International Codes as well.

CODE ADOPTION

The IRC and the family of International Codes are referred to as *model codes*, nationally recognized construction regulations that serve as models for local ordinances and are maintained and updated through an open process of code development. Such a process relies on the participation of experts and all interested parties in the various fields of design, construction, manufacturing, government, and code administration. These model codes are updated on three-year cycles to recognize new and developing technology, materials, and methods of construction. Changes in the codes are also often in response to natural or human-made disasters involving the loss of lives or the destruction of property.

Adoption of the IRC

The IRC becomes an enforceable regulation through the legal proceedings of the governmental jurisdiction. The adopting ordinance references the edition and title of the IRC and provides other necessary factual information (Figure 2-1). Typically the legislation also includes the purpose, scope, and effective date for the ordinance. As part of the adoption process, the government authority must also provide local information for insertion into code text, including the name of the jurisdiction, design criteria (see Chapter 4), and building sewer depths (see Chapter 13). **[Ref. R101.1, Table R301.2(1), and P2603.6.1]**

ORDINANCE NO. 000-06
Residential Building Regulations

An ordinance of the city adopting the 2009 edition of the International Residential Code, regulating and governing the construction, alteration, movement, enlargement, replacement, repair, equipment, location, removal and demolition of detached one and two family dwellings and townhouses not more than three stories in height with separate means of egress and their accessory structures in the city providing for the issuance of permits and collection of fees therefore; repealing Ordinance No. 000-03 of the city and all other ordinances and parts of the ordinances in conflict therewith.

The city council does ordain as follows:

I. Except as hereafter modified, that the International Residential Code, 2009 Edition, including Appendix chapters _____, as published by the International Code Council, is hereby specifically adopted by reference as the residential code of the city. The provisions of said residential code shall be controlling in the design, construction, quality of materials, erection, installation, addition, alteration, repair, location, relocation, replacement, removal, demolition, use and maintenance of detached one and two family dwellings and townhouses not more than three stories in height with separate means of egress and their accessory structures in the city and in all matters covered by said residential code within the city.

II. The following sections are hereby revised:
Section R101.1. Insert: [NAME OF JURISDICTION]
Table R301.2(1) Insert: [APPROPRIATE DESIGN CRITERIA]
Section P2603.6.1 Insert: [NUMBER OF INCHES IN TWO LOCATIONS]

III. That this ordinance and the rules, regulations, provisions, requirements, orders and matters established and adopted hereby shall take effect and be in full force and effect upon the date of its final passage and adoption.

FIGURE 2-1 Sample ordinance adopting the IRC

Amending the IRC

Advantages to the adoption of model codes include consistency, uniformity, and correlation among construction codes and across jurisdictional boundaries. Such uniformity benefits designers and builders as well as government officials in charge of administering the codes. While states, cities, counties, and similar governmental jurisdictions typically do not have the resources to develop and maintain comprehensive construction codes, jurisdictions do have the ability to modify the model code through amendments placed in the adopting ordinance. Excessive local amendments to adopted model codes may be contrary to the goals of consistency and may offset the advantages and legal defensibility of nationally recognized standards. Still, some modification of the IRC may be necessary or desirable for any number of reasons.

Though model codes anticipate location and climate differences, amendments to them may be influenced by unique characteristics and conditions, such as geographic location, weather, topography, flooding, soil properties, and water tables. There also may be considerations of political situations, local traditions or customs, compatibility issues with other state or local laws, or the existence of unique housing stock, such as in historic districts. Lengthy amendments to the IRC are not uncommon. It is important to recognize that these amendments must be legally instituted with care through the adopting ordinance and laws of the jurisdiction. Otherwise, actions to enforce provisions not contained in the IRC or to waive certain requirements that are in the model code, in addition to creating inconsistency in the application of the code, will not be in accordance with the law.

Appendices

The appendices are developed in much the same manner as the main body of the model code. However, the appendix information is judged outside the scope and purpose of the code at the time of code publication. Many times an appendix offers supplemental information, alternative methods, or recommended procedures. The information may also be specialized and applicable or of interest to only a limited number of jurisdictions. While an appendix may provide some guidelines or examples of recommended practices or assist in the determination of alternative materials or methods, it will have no legal status and cannot be enforced until it is specifically recognized in the adopting legislation. Appendix chapters or portions of such chapters that gain general acceptance over time are sometimes moved into the main body of the model code through the code development process. [**Ref. R102.5**]

Local and state laws

The IRC is not meant to nullify any local, state, or federal law, and in many cases, such other laws supersede provisions found in the building code. Careful consideration should be given to the interaction of the residential code with other codes and laws of the jurisdiction and the state during the process of adoption. Zoning ordinances, for example, may include more restrictive provisions on distances to property lines and between structures, permissible heights and areas of buildings, the number of buildings on a lot, and the specific conditions of use and occupancy of residences and accessory buildings. Other local ordinances impacting the residential code and construction may include those regulating public streets, storm water management, erosion control, public and private sewers, and private wells. State laws may override the building codes in matters related to energy conservation, manufactured and modular housing, and accessibility for persons with disabilities. State law also often determines the circumstances under which a licensed engineer or architect is required, and sets the licensing regulations for these design professionals (Figure 2-2). [Ref. R102.2]

FIGURE 2-2 Engineer's seal

AUTHORITY

The IRC establishes a department of building safety, commonly referred to as the building department, and designates the officer in charge of the administration and enforcement of the code as the building official. The appointing authority of the jurisdiction appoints the building official, though jurisdiction job titles may vary. The building official in turn assigns some degree of decision-making authority to deputies, plans examiners, inspectors, permit technicians, and other employees. The role of a building official in protecting public safety is complex and challenging. It follows that the position demands skills, knowledge, and abilities to not only fulfill the duties, but elevate the credibility of the department in the eyes of the public.

Authority and duties of the building official

The IRC charges the building official with enforcing the provisions of the code and assigns broad authority and discretion to do so. With discretion comes the responsibility to make decisions in keeping with the intent of the IRC. The building official does not have authority to waive code requirements. In the same way, the building official has no authority to require more than the code stipulates. The IRC also authorizes the building official to develop policies and procedures for consistent application of the code provisions. [Ref. R104.1]

In effectively performing the duties listed in the code, the building official must also have an understanding of the legal aspects of code administration. While given broad authority for enforcement, including the issuance of notices and orders, the building official must also recognize the rights of due process afforded to the public. Equally important to the building department in securing safe buildings for the community is to build the public trust through communication, respect, and fairness, so that the department is viewed as a resource rather than an adversary.

You Should Know

"The building official is hereby authorized and directed to enforce the provisions of this code." ●
—International Residential Code.

Interpretations

Many code provisions are clear and easily understood. The dimensions for stair rise and run, handrail height, and spacing for openings in guards, for example, are specific and objectively measurable. Such provisions are referred to as *prescriptive*—a clear set of rules to follow to gain compliance. The shape of other than round handrails, on the other hand, is less clearly defined. While the code sets some parameters for dimensions of these handrails, it also permits any handrail that provides equivalent graspability.

The term "equivalent graspability" is a performance requirement, meaning that an element must function to satisfy certain acceptable criteria. Determination of compliance requires some level of judgment on the part of the building official. Though the IRC intends to be largely prescriptive in nature, it purposely offers performance criteria as well, to allow flexibility in design and construction and to not favor certain materials or methods over any other.

Alternative methods and materials and evaluation service reports

The IRC is specific in its intention to not exclude the use of any material or method of construction, even if such methods are not specifically described by the code, subject to approval by the building official. The building official has an obligation, as instructed by the code, to approve such alternatives where he or she finds that the proposed material or construction meets the intent of the IRC and is equivalent to the code provisions. With modern technology advancing at a record pace, new and innovative building products are continuously introduced to the market on a global scale. Reports issued by the International Code Council's Evaluation Service (ICC-ES) are valuable resources in verifying performance equal to the code requirements (see Figure 2-3). In the absence of ICC-ES evaluation reports and where insufficient data or documentation exists, the building official may require that tests be performed by an approved agency to demonstrate compliance with the code. Also, compliance with the specific performance-based provisions of the referenced International Codes satisfies the IRC requirements. The performance and alternative-methods provisions of the IRC and the use of the ICC Evaluation Reports provide for flexibility and encourage innovative and new materials, design, and construction while protecting the public safety. All published ICC Evaluation Reports are available free of charge and can be accessed online at www.icc-es.org. [**Ref. R104.10 and R104.11**]

PERMITS

Except for a short list of work of a minor nature, any construction requires a permit before work begins, including work for the relocation or demolition of buildings. Work exempt from permits must still comply with the applicable IRC requirements. [**Ref. R105**]

 **ES REPORT™**

ESR-4802

Issued September 1, 2009
This report is subject to re-examination in one year.

ICC Evaluation Service, Inc.
www.icc-es.org

Business/Regional Office # 5360 Workman Mill Road, Whittier, California 90601 # (562) 699-0543
Regional Office # 900 Montclair Road, Suite A, Birmingham, Alabama 35213 # (205) 599-9800
Regional Office # 4051 West Flossmoor Road, Country Club Hills, Illinois 60478 # (708) 799-2305

DIVISION: 07—THERMAL AND MOISTURE PROTECTION
Section: 07410—Metal Roof and Wall Panels

REPORT HOLDER:

ACME CUSTOM-BILT PANELS
52380 FLOWER STREET
CHICO, MONTANA 43820
(808) 664-1512
www.custombiltpanels.com

EVALUATION SUBJECT:

CUSTOM-BILT STANDING SEAM METAL ROOF PANELS:
CB-150

1.0 EVALUATION SCOPE

Compliance with the following codes:

- 2009 *International Building Code®* (IBC)
- 2009 *International Residential Code®* (IRC)

Properties evaluated:

- Weather resistance
- Fire classification
- Wind uplift resistance

2.0 USES

Custom-Bilt Standing Seam Metal Roof Panels are steel panels complying with IBC Section 1507.4 and IRC Section R905.10. The panels are recognized for use as Class A roof coverings when installed in accordance with this report.

3.0 DESCRIPTION

3.1 Roofing Panels:

Custom-Bilt standing seam roof panels are fabricated in steel and are available in the CB-150 and SL-1750 profiles. The panels are roll-formed at the jobsite to provide the standing seams between panels. See Figures 1 and 3 for panel profiles.

The standing seam roof panels are roll-formed from minimum No. 24 gage [0.024 inch thick (0.61 mm)] cold-formed sheet steel. The steel conforms to ASTM A 792, with an aluminum-zinc alloy coating designation of AZ50.

3.2 Decking:

Solid or closely fitted decking must be minimum $^{15}/_{32}$-inch-thick (11.9 mm) wood structural panel or lumber sheathing, complying with IBC Section 2304.7.2 or IRC Section R803, as applicable.

4.0 INSTALLATION

4.1 General:

Installation of the Custom-Bilt Standing Seam Roof Panels must be in accordance with this report, Section 1507.4 of the IBC or Section R905.10 of the IRC, and the manufacturer's

published installation instructions. The manufacturer's installation instructions must be available at the jobsite at all times during installation.

The roof panels must be installed on solid or closely fitted decking, as specified in Section 3.2. Accessories such as gutters, drip angles, fascias, ridge caps, window or gable trim, valley and hip flashings, etc., are fabricated to suit each job condition. Details must be submitted to the code official for each installation.

4.2 Roof Panel Installation:

4.2.1 CB-150: The CB-150 roof panels are installed on roofs having a minimum slope of 2:12 (17 percent). The roof panels are installed over the optional underlayment and secured to the sheathing with the panel clip. The clips are located at each panel rib side lap spaced 6 inches (152 mm) from all ends and at a maximum of 4 feet (1.22 m) on center along the length of the rib, and fastened with a minimum of two No. 10 by 1-inch pan head corrosion-resistant screws. The panel ribs are mechanically seamed twice, each pass at 90 degrees, resulting in a double-locking fold.

4.3 Fire Classification:

The steel panels are considered Class A roof coverings in accordance with the exception to IBC Section 1505.2 and IRC Section R902.1.

4.4 Wind Uplift Resistance:

The systems described in Section 3.0 and installed in accordance with Sections 4.1 and 4.2 have an allowable wind uplift resistance of 45 pounds per square foot (2.15 kPa).

5.0 CONDITIONS OF USE

The standing seam metal roof panels described in this report comply with, or are suitable alternatives to what is specified in, those codes listed in Section 1.0 of this report, subject to the following conditions:

5.1 Installation must comply with this report, the applicable code, and the manufacturer's published installation instructions. If there is a conflict between this report and the manufacturer's published installation instructions, this report governs.

5.2 The required design wind loads must be determined for each project. Wind uplift pressure on any roof area must not exceed 45 pounds per square foot (2.15 kPa).

6.0 EVIDENCE SUBMITTED

Data in accordance with the ICC-ES Acceptance Criteria for Metal Roof Coverings (AC166), dated October 2007.

7.0 IDENTIFICATION

Each standing seam metal roof panel is identified with a label bearing the product name, the material type and gage, the Acme Custom-Bilt Panels name and address, and the evaluation report number (ESR-4802).

FIGURE 2-3 Sample ICC-ES Evaluation Report

Permit application

The owner or authorized agent must make application for a permit on a form furnished by the building department. In addition to providing a legal description of the property, the permit application must include the description of the work, the valuation of the proposed work, the use of the building, and the applicant's signature (Figure 2-4).

Plans and specifications

Construction drawings and other submittal documents must accompany the permit application and be of sufficient detail and clarity to verify compliance with the code. The code also requires a site plan showing all new and existing structures with distances to lot lines. The extent of construction documents varies with the complexity and scope of the project. The building official is authorized to waive submittal documents for work of a minor nature, provided that code compliance can be verified by other means. Local or state laws determine requirements for a registered design professional to prepare the construction documents. In the case of any special conditions as determined by the building official, such as the sizing of a steel beam or the support for a concentrated load, the building official is also authorized to require plans to be prepared by a registered architect or engineer. **[Ref. R106]**

A detailed review of plans and specifications is necessary to verify that the design complies with the code, thereby avoiding costly modifications during the course of construction. Jurisdictional policies differ in the handling of incomplete or incorrect plans. Depending on the complexity of the project and the significance of errors or omissions, the plans examiner, on behalf of the building official, may furnish comments or a list of code requirements to the applicant, may request supplemental information from the applicant, or may reject the plans and require submittal of revised documents (Figures 2-5 and 2-6).

Fees

The jurisdiction establishes a schedule of fees at a level sufficient to offset the costs of providing the associated services to the public, including administration, plan review, and inspection. Permit fees are often based on the total value of the work included in the scope of the permit, though other methods may be used. In any case, the building official should develop an equitable and consistent procedure for assessing fees related to the permit process (Figure 2-7). **[Ref. R108]**

Permit issuance

In the language of the code, the building official must review the application and construction documents within a reasonable time and, when approved, issue the permit as soon as is practicable (Figure 2-8). The length of time considered reasonable will vary based on several factors, such as the complexity of the project and the completeness of the construction documents. The intent of the code is to allow sufficient time for a thorough review of plans to determine code compliance and to issue

**Building Department
City Of**

Application for Permit

Project Address _____

Parcel _____ Lot _____ Subdivision _____

_____ Zone _____ Flood Zone _____

Owner _____ Phone _____

Address _____

Contractor _____ Phone _____

Address _____

Project Type ☐ New ☐ Addition ☐ Alteration ☐ Repair ☐ Demolition

Proposed Use _____

Work Description _____

_____

_____

_____

_____

_____

Total Square Ft. _____ Valuation $ _____ Fee _____

I hereby certify that I have read and examined this document and know the same to be true and correct. All provisions of laws and ordinances governing this type of work will be complied with whether specified herein or not. I further certify that I am the owner or the owner's authorized agent and that the proposed work is authorized by the owner. I understand that work shall not begin until the permit is issued by this department, that I am responsible for calling for all required inspections, that work shall be accessible for Inspection, that a final inspection, approval and Certificate of Occupancy are required prior to occupying this building. Fees are non-refundable, except when the permit and construction are cancelled before work begins, in which case the applicant may apply for a partial refund in accordance with the refund policy. This permit application is only for the work described above. Every permit issued shall become invalid unless the work authorized by such permit is commenced within 180 days after its issuance, or if the work authorized by such permit is suspended or abandoned for a period of 180 days after the time the work is commenced.

Applicant _____ Signature _____

Address _____ Phone _____

Amount Paid _____ Date _____ Received by _____

FIGURE 2-4 Application for permit

Residential Plan Review Checklist
2009 International Residential Code

Foundations & Concrete—IRC Chapter 4	**Section**
☐ Lots graded to drain surface water away from foundation walls ≥ 6 inches fall within the first 10 feet	R401.3
☐ Concrete minimum specified compressive strength: • Footings, interior slabs: 2500 psi • Walls exposed to weather: 3000 psi • Garage slabs and exterior slabs 3500 air-entrained	Table R402.2
☐ Footings supported on undisturbed natural soils at least 12 inches below undisturbed ground or on engineered fill. Footing sizes: • Spread • Trench • Mat, pier, post, fireplace	R401.2, R403.1
☐ Foundation walls extend above the finished grade ≥ 6 inches (4 inches where masonry veneer is used)	R404.1.6
☐ Foundation anchor bolts ≥ ½" diameter extending ≥ 7 in. into masonry or concrete, maximum 6 ft OC and within 12" of ends	R403.1.6
☐ Concrete slab-on-ground floors ≥ 3.5 in. thick	R506.1
☐ Approved vapor retarder under slab	R506.2.3
☐ Approved drainage pipe at or below the area to be protected on a ≥ 2 inches of ¾ inch minimum washed crushed rock and covered with ≥ 6 inches of the same material.	R405.1
☐ Basement walls dampproofed (waterproofed if high water table)	R406.1, R406.2
Floors—IRC Chapter 5	
☐ Spans for wood floor joists	Tables R502.3.1 (1) & (2)
☐ Spans of girders	Tables R502.5 (1) & (2)
☐ End bearing of joist beam or girder: • ≥ 1.5 inches on wood or metal • ≥ 3 inches on masonry or concrete • Or supported by approved joist hangers	R502.6
☐ Joists framing from opposite sides over bearing support shall lap ≥ 3 in.	R502.6.1
☐ Manufactured floor I-joist shall be installed in accordance with manufacturer's instructions.	
☐ Engineered floor truss design drawings and location drawing: • Hanger type & location • Approved connections • Bracing per engineered truss design drawings • Trusses shall not be cut, notched, spliced or altered	R502.11
☐ Draftstops when usable space above and below the concealed space of a floor/ceiling assembly with area ≤ 1000 square feet and approximately equal areas	R502.12, R302.12

FIGURE 2-5 Plan review checklist

the permit in a timely manner to avoid unnecessary delays in the construction schedule, resulting in hardship to the applicant. A copy of the permit and the approved construction documents must be kept on the jobsite until completion of the project.

REVIEWED FOR CODE COMPLIANCE

FIGURE 2-6 Approved plans stamp

City Building Department
PERMIT FEE SCHEDULE
As adopted by city resolution number 00-0000 effective ___/___/___

Building Department
City Of

TOTAL VALUATION	PERMIT FEE
$1 to $500	$24.00
$501 to $2,000	$24.00 for the first $500; plus $3 for each additional $100 or fraction thereof, to and including $2,000
$2,001 to $40,000	$69.00 for the first $2,000; plus $11 for each additional $1,000 or fraction thereof, to and including $40,000
$40,001 to $100,000	$487 for the first $40,000; plus $9 for each additional $1,000 or fraction thereof, to and including $100,000
$100,001 to $500,000	$1,027 for the first $100,000; plus $7 for each additional $1,000 or fraction thereof, to and including $500,000
$500,001 to $1,000,000	$3,827 for the first $500,000; plus $5 for each additional $1,000 or fraction thereof, to and including $1,000,000
$1,000,001 to $5,000,000	$6,327 for the first $1,000,000; plus $3 for each additional $1,000 or fraction thereof, to and including $5,000,000
$5,000,001 and up	$18,327 for the first $ 5,000,000; plus $1 for each additional $1,000 or fraction thereof

The applicant shall state the valuation of the proposed work at the time of application for a permit. The building official shall make the final determination of valuation in accordance with established department guidelines and published cost data such as ICC Building Valuation Data. Building permit valuation shall include total value of the work for which a permit is being issued, such as electrical, gas, mechanical, plumbing equipment and other permanent systems, including materials and labor.

Refunds shall be in accordance with the refund policy established by the building official in accordance with IRC Section R108.5.

FIGURE 2-7 Permit fee schedule

Code Basics

Permit-holder responsibilities:

- Permit on jobsite
- Approved plans on jobsite
- Call for inspection
- Provide access for inspection ●

INSPECTIONS

Inspection of the work by qualified personnel at various stages throughout the construction process is essential to verify compliance with the code and the approved plans. It is the responsibility of the permit holder or agent to call for the required inspections before work is concealed and to provide access to such work. If inspection reveals work that does not comply with the code, the inspector notifies the permit holder or agent of the deficiencies requiring correction. The inspector must approve the corrected portions of the work before they are concealed. When work is satisfactory, the inspector typically indicates approval with an inspection sticker or tag (Figure 2-9) or by signing a record-of-inspections card authorizing work to proceed (Figure 2-10). **[Ref. R109]**

City of - - - -

Building Department
City Of

Notice
Building Permit

Permit No. 00-0000

Has been issued to

Owner ______________________

Contractor ______________________

For (Work Description) ______________________

____

____

At (Project Address) ______________________

(Date) ______________________ (Building Official Signature) ______________________
Issue Date **Building Official**

The issuance or granting of a permit shall not be construed to be a permit for, or an approval of, any violation of any of the provisions of the city building code or of any other ordinance of the jurisdiction.

A copy of this building permit shall be kept on the site of the work until the completion of the project.

The permit holder or authorized agent is responsible for calling for all required inspections before work is covered and for providing access to the work. Do not occupy this building, or portion of building as described, until final inspection, approval and issuance of the certificate of occupancy.

This permit expires ___/___/___

City Building Department
(000) 000-0000

FIGURE 2-8 Issued building permit card

Department of Building Safety
Phone (###) 555-4567

INSPECTION APPROVED

- ☐ Building
- ☐ Electrical
- ☐ Plumbing
- ☐ Mechanical

Description: ____________________

Comments ____________________

____

____

Date: ____________________

Inspector: ____________________

FIGURE 2-9 Inspection approval tag

Required inspections

The IRC specifically requires certain inspections during the course of construction when applicable, as set forth in Table 2-1. [Ref. R109]

Other inspections

The building official is authorized to require additional inspections as may be necessary to ensure compliance with the code. Many jurisdictions perform insulation and energy inspections to verify compliance with energy conservation provisions. Inspections may also be desirable to verify the installation of unusual, special, or engineered components or systems. For example, a cast-in-place or precast structural concrete floor or deck (Figure 2-11) is an engineered structure, the design of which is outside the scope of the IRC. The inspector must verify dimensions, materials, anchorage, reinforcing, and other details to ensure conformance to the engineered drawings.

Certificate of occupancy

The IRC requires issuance of a certificate of occupancy (Figure 2-12), indicating that the work has passed final inspection before a dwelling unit, or the portion of the dwelling unit covered by the permit, can be occupied. [Ref. R110]

Inspection Record

Building Department
City Of

Permit No. _____

Name _____

Address _____

INSPECTOR SHALL SIGN ALL SPACES WHICH APPLY TO THIS JOB

Inspection Category	Date	Comment	Inspector
Foundation Inspection:			
Setbacks, Footings			
Under Slab Inspections:			
Plumbing			
Mechanical			
Electrical			
Utility Inspections:			
Electrical Service			
Gas Piping/Air Test			
Rough-In Inspections:			
Plumbing			
Mechanical (HVAC)			
Electrical			
Framing			

DO NOT COVER WORK
UNTIL IT IS INSPECTED, APPROVED AND ABOVE SPACES ARE SIGNED

Inspection Category	Date	Comment	Inspector
Final Inspection			

DO NOT OCCUPY
Final Inspection Approval and Certificate of Occupancy
issued by the building department are required before occupying this building

FIGURE 2-10 Jobsite inspection record card

TABLE 2-1 Required inspections

Inspection:	Conducted when:	Inspect:
Foundation	• excavation complete • forms set • reinforcing in place • before any concrete placed	• setback from lot lines • dimensions • reinforcing • suitable soil or base materials • vegetation, loose soil, and debris removed • sufficient depth for frost protection • concrete mix specifications
Floodplain	• the lowest floor is in place (applies to only designated flood hazard areas)	• elevation certificate prepared and sealed by a registered design professional before work proceeds above
Plumbing, mechanical, gas, and electrical rough-in inspections	• underground work complete • rough-in stage complete prior to rough frame inspection • utility connections	• materials, fittings, and methods • work properly supported and protected • pressure testing of piping systems
Frame and masonry	• plumbing, mechanical, and electrical rough inspections approved • framing and masonry complete	• size, spacing, connection, and continuity of all structural members • load path from roof to foundation • engineered components • draftstopping and fireblocking • stair rise and run • locations and dimensions of emergency-escape openings

Fire resistance rated construction	• gypsum board fire resistance rated assemblies
	  • between dwelling units
	  • exterior walls due to fire separation distance to lot lines
		• gypsum materials
		• fastener type, size, and spacing conform to the approved fire resistance rated assembly details

Final	• work under permit complete and ready for occupancy	• comfort heating, cooling, and ventilation equipment and systems • plumbing fixtures and systems • electrical circuits, devices, and fixtures • stairs, railings, landings, and other means of egress components • smoke alarms • emergency-escape and -rescue openings • other items regulated by the code

BOARD OF APPEALS

The administration chapter of the IRC creates authority and duties for the building official but intends that actions in enforcing the code be reasonable, and also clearly expresses rights of due process for the public. Such is the case in the right to appeal an order, decision, or determination of the building official. Any aggrieved party with a material interest in the decision of the building official may apply for redress to the board of appeals (Figure 2-13). The governing body, such as the city council, appoints the board members, who are qualified by experience and training to hear and rule on matters related to building construction.

FIGURE 2-11 Precast concrete floor

Certificate of Occupancy

Jurisdiction of _____

This structure or portion of structure as described below has been inspected for compliance with the International Residential Code (IRC) and is hereby issued a certificate of occupancy.

Description of applicable
portion of structure _____

Address of the structure _____

Owner's name _____

Owner's address _____

Building permit
number _____

Code edition **2009 IRC**

Sprinkler system
required ☐ yes ☐ no

Sprinkler system
installed ☐ yes ☐ no

Special Conditions _____
_____

Building Official _____

FIGURE 2-12 Required information for certificate of occupancy

Department of Building Safety
Application for Appeal
Board of Appeals

**Building Department
City Of**

Project address _____

Use of structure _____

Description of work _____

Owner's name _____ Phone _____

Owner's address _____

In accordance with the provisions of Section R112 of the *International Residential Code for One and Two Family Dwellings* (IRC), I hereby appeal to the Board of Appeals the determination made by the building official relative to the interpretation of Section _____, in order that I might construct the above structure or portion thereof as proposed and shown on the attachments.

Appellant is advised to submit any documentation in support of the appeal. An application for appeal shall be based on a claim that the true intent of the code has been incorrectly interpreted, the provisions of the code do not fully apply, or an equally good or better form of construction is proposed. The board has no authority to waive requirements of the code. Appellant and any interested party may appear to present reasons for granting an appeal at the time of the scheduled meeting.

_____
Signature of owner or appellant   Date

Meeting date _____

Time _____

Location _____

FIGURE 2-13 Application for appeal

The intent is to put in place an objective and knowledgeable group of citizens to review the decisions of the building official and consider the merits of any appeal.

The IRC limits the basis for appeal to matters pertaining to the code requirements. The appellant must claim that the building official has erred in interpreting the code or has wrongly applied a code section. The other basis for appeal is that the appellant considers a proposed alternative to be equal to the code requirements. The IRC does not permit the filing of an appeal seeking a variance or a waiver, and the board has no authority to waive code requirements. [Ref. R112]

Code Basics

Basis for appeal:

- The code has been interpreted incorrectly.
- The code does not apply.
- An equal alternative is proposed. ●

PART

II

Site Development

Chapter 3: Site Preparation

Site Preparation

In preparation for constructing buildings on a property, the builder must consider a number of factors related to code requirements. The buildings must be located according to the approved site plan to meet the requirements of the *International Residential Code* (IRC) and any applicable local ordinances. The soil must be suitable for the support of the building and is factored into the design of the foundations. And the building must be elevated sufficiently and the site graded to provide surface drainage away from the building. The plans examiner considers these factors when checking the construction drawings and site plan, but the inspector will be responsible for verifying the requirements at the jobsite (Figure 3-1).

FIGURE 3-1 Sitework

LOCATION ON PROPERTY

The IRC regulates a building's location on the property primarily to guard against the spread of fire. The code is concerned with not only protecting the new building on the property being developed, but preventing the spread of fire to adjacent buildings. Structural considerations also play a part in locating buildings on a lot. The code regulates distances between the structure and adjacent steep slopes to protect the integrity of the foundation. Local zoning or other ordinances may be more restrictive in regulating the location, height, and area of buildings on properties.

Fire separation distance

By definition, fire separation distance is measured from the face of the building to the lot line, centerline of a street or alley, or to an imaginary line between two buildings. However, for all practical purposes, fire separation distance typically will be of concern only when measured to the lot line. No separation distance or fire resistance rating is required for opposing walls of detached structures on the same lot. Fire separation distance is measured at a right angle to the face of the exterior wall. [Ref. R202]

Separation distance from lot line

Exterior walls of dwellings, garages, and accessory buildings must maintain a minimum 5 feet separation distance from the lot line, measured perpendicular to the wall, or be protected for fire. Walls less than 5 feet from the lot line require a fire resistance rating of one hour when exposed to fire from either side of the wall. (Fire resistant rated construction is discussed further in Chapter 9.) There is nothing in the code to prohibit buildings from sitting on the property line, provided they meet the fire resistance requirements and have no roof overhang or openings in the exterior wall (Figures 3-2 and 3-3).

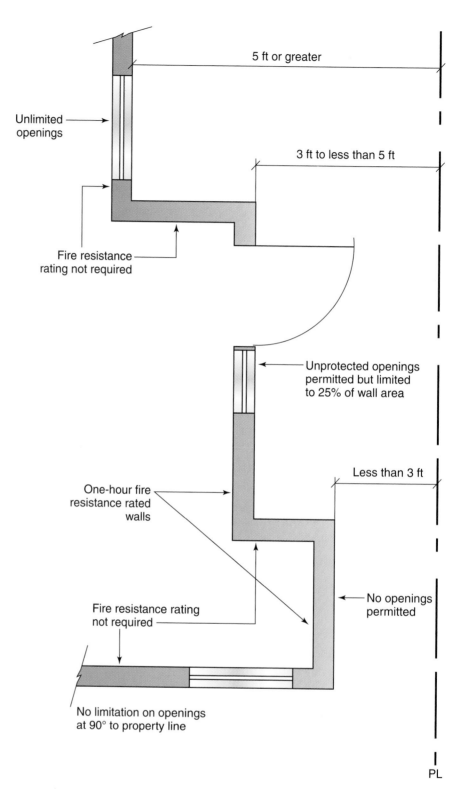

FIGURE 3-2 Exterior wall location and fire separation distance

The IRC also regulates projections and openings within this 5-foot measurement to a lot line. Openings, typically windows or doors, are not permitted less than 3 feet from the lot line. When they are more than 3 feet but less than 5 feet from the lot line, the code limits the openings to 25 percent of the wall area. Foundation vents for ventilation of

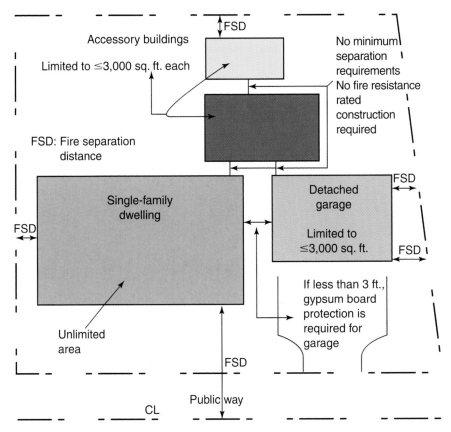

Exterior walls of dwellings and accessory buildings on same lot

FIGURE 3-3 Measuring fire separation distance

TABLE 3-1 Exterior walls

Exterior Wall Element		Minimum Fire Resistance Rating (hours)	Minimum Fire Separation Distance (ft.)
Walls	Fire resistance rated	1 (with exposure from both sides)	<5
	Not fire resistance rated	0	5
Projections	Fire resistance rated	1 (on the underside)	2
	Not fire resistance rated	0	5
Openings	Not allowed	N/A	<3
	25% maximum of wall area	0	3
	Unlimited	0	5
Penetrations	All	Comply with Section R302.4	<5
		None required	5

underfloor spaces are an exception to this rule and are permitted. Projections, typically roof overhangs, must have one-hour fire protection on the underside when less than 5 feet from the lot line and cannot project closer than 2 feet from the lot line. An exception permits a garage located within 2 feet of the lot line to have a 4-inch roof eave projection (Table 3-1 and Figure 3-4). **[Ref. R302.1]**

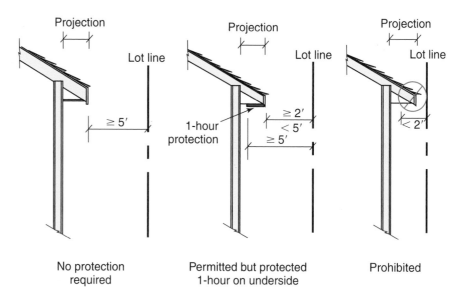

FIGURE 3-4 Roof projections at property line

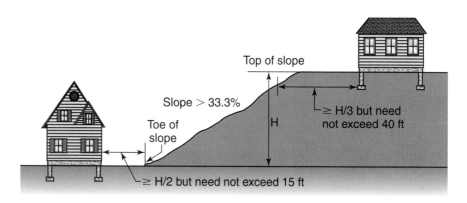

FIGURE 3-5 Foundations adjacent to slopes

Location of foundations adjacent to slopes

Where slopes are steeper than 33.3 percent (4 inches per foot), foundations must be located a sufficient distance away from the slope to protect the integrity of the structure and provide adequate lateral support to the footing. The clearance distance is based on the height of the slope. For a building located adjacent to the top of the slope (descending), the minimum distance is the height divided by 3, but does not need to exceed 40 feet. For a building located adjacent to the bottom of the slope (ascending), the minimum clearance is the height divided by 2, but does not need to exceed 15 feet. The code gives the building official the authority to approve alternate setbacks with lesser distances to slopes based on a design by a qualified engineer taking all site conditions into consideration (Figure 3-5). [Ref. R403.1.7]

SITE PREPARATION

Regulation of site preparation activities related to construction of buildings under the IRC varies based on geographic location and local or site-specific conditions. The code is basically concerned with two things: soil characteristics related to the support and stability of foundations, and grading to provide surface drainage away from foundations. Additionally, construction in flood hazard areas must comply with the elevation and design requirements of the IRC or local floodplain regulations. There may also be local or state laws that require grading permits and regulate erosion control, storm water management, and soil conservation measures. A number of other factors may affect site preparation and building design, including high water tables and sloped sites.

General requirements

Preparation of the site for construction includes stripping of vegetation and topsoil, grading to the rough contours if necessary, and excavation for basements and foundations. The IRC requires that all exterior footings be placed at least 12 inches below the undisturbed ground level and be protected against frost where applicable. Footings must bear on undisturbed natural soil or compacted engineered fill (covered later in this chapter under "Fill"). The code also prescribes suitable base requirements for basement and garage floors, other slabs on grade, and the base for crawl spaces. In all cases, the ground must be stripped of vegetation and organic material. The base for concrete floor slabs within the perimeter walls must be of suitable materials and compacted to prevent settlement. The thickness of compacted fill material below slabs is generally limited to 24 inches for clean sand or gravel and 8 inches for soil unless otherwise approved by the building official (Figure 3-6). **[Ref. R403.1, R408.5, R506.2]**

Soil properties

The designer or builder must carefully consider soil properties not only for adequate support of the foundation but also for stability to prevent future damage to the structure. Based on experience and known local soil conditions, the building official will often permit design based on a presumptive load-bearing value without soil testing or a geotechnical report. Typically, the presumed load-bearing value will range from 1500

FIGURE 3-6 Excavation

to 3000 pounds per square foot (psf) based on local soil conditions and according to the values in Table 3-2. The building official may assume conservative values based on the average or the lowest soil characteristics likely to be encountered on a site. Soil type is verified at the time of footing inspection. If found to be of a poorer grade than the presumed value, testing or mitigation is required prior to placing concrete footings. The builder always has the option of providing the results of soil testing in a geotechnical report in order to use a higher load-bearing value than would otherwise be presumed. [Ref. R401.4.1]

Where available data indicates that the soil may not be suitable for the foundation design, the building official is authorized to require a geotechnical

TABLE 3-2 Presumptive load-bearing values and properties of soils

Unified Soil Classification System Symbol	Soil Description	Load-Bearing Pressure (psf)	Drainage	Frost Heave Potential	Volume Change Potential Expansion
GW	Well-graded gravels, gravel-sand mixtures, little or no fines	3000	Good	Low	Low
GP	Poorly graded gravels or gravel-sand mixtures, little or no fines	3000	Good	Low	Low
SW	Well-graded sands, gravelly sands, little or no fines	2000	Good	Low	Low
SP	Poorly graded sands or gravelly sands, little or no fines	2000	Good	Low	Low
GM	Silty gravels, gravel-sand-silt mixtures	2000	Good	Medium	Low
SM	Silty sand, sand-silt mixtures	2000	Good	Medium	Low
GC	Clayey gravels, gravel-sand-clay mixtures	2000	Medium	Medium	Low
SC	Clayey sands, sand-clay mixture	2000	Medium	Medium	Low
ML	Inorganic silts and very fine sands, rock flour, silty or clayey fine sands, or clayey silts with slight plasticity	1500	Medium	High	Low
CL	Inorganic clays of low to medium plasticity, gravelly clays, sandy clays, silty clays, lean clays	1500	Medium	Medium	Medium to low
CH	Inorganic clays of high plasticity, fat clays	1500	Poor	Medium	High
MH	Inorganic silts, micaceous or diatomaceous fine sandy or silty soils, elastic silts	1500	Poor	High	High

[Ref. Tables R401.4.1 and R405.1]

evaluation and report prepared by an approved agency or registered design professional. Expansive, compressible, or shifting soils have the potential to damage the structure. Highly organic soils (laden with decayed material from plants and animals), such as organic clays, organic silts, and peat, are not included in Table 3-2 and are outside the scope of foundation design under the IRC. In addition to organic materials, certain inorganic clays and silts are also highly expansive. Such soils expand when wet and contract as they dry, exerting significant pressures against the footing and foundation and thereby causing shifting or differential settlement that could result in structural failure. Expansive soil conditions require an engineered foundation design in accordance with the *International Building Code* (IBC). In some cases it may be possible to remove unsuitable shifting or compressible soils from the building site and replace them with approved fill to stabilize the soil below and around foundations. Under these conditions, the IRC permits a prescriptive foundation design without a full geotechnical evaluation. [Ref. 401.4]

Fill

Overexcavation to remove unsuitable soils or the addition of material to raise the elevation of the footings above the level of the natural undisturbed soil requires engineered fill material to support the footings and foundation. A registered design professional is responsible for the design and placement of the fill material in accordance with accepted engineering practice. The engineered fill must be installed and tested in conformance with the design requirements. Fill materials are typically sand, crushed rock, clean gravel, or a mix of granular materials. Fill material may contain finer particles that fill voids and help bind the larger elements together. Materials with rounded edges such as river rock or pea gravel are not usually considered suitable for structural fill. The engineer's design specifies the maximum thickness of each layer of fill, called a *lift,* prior to mechanical compaction. A technician tests the compacted fill to verify that it meets the minimum compaction and design specifications. Builders should also exercise care during the backfill of foundations with suitable fill materials to provide adequate drainage and to prevent damage to the foundation. [Ref. R106.1, R401.2]

STORM DRAINAGE

The IRC prescribes methods to direct surface water away from the foundation to an approved location. Water held against the foundation leads to wet or damp basements or crawl spaces and over time can cause damage to construction materials both inside and outside the structure. Mold thrives in such moist environments, contributing to an unhealthy living environment. In addition, water saturation of the soils adjacent to foundations increases the lateral pressure against the structure. Proper design of surface drainage also prevents nuisance ponding on the lot and possible flooding of structures during periods of heavy rain.

The IRC lends some discretion to the building official in determining alternate methods for adequate drainage. Department policy for verifying proper surface drainage on properties will likely vary depending on geographic location, permeability of soils, and local history of damage and nuisances created by inadequate drainage. The building official is authorized to require submittal documentation sufficient to demonstrate compliance with the code. If deemed necessary, this may entail a detailed drainage plan with existing and proposed topographic contours, elevations, points of discharge, and any containment features. The building official may require that a registered design professional prepare such drainage plans. In many cases, a drainage plan is already established as part of the master plan for the entire housing development, and additional plans are not necessary. Other jurisdictions may require only some indication of the direction of drainage flow on the required site plan or may verify drainage on site visually without measurement at the time of inspection (Figure 3-7).

The IRC is most concerned with drainage in the immediate vicinity of the structure. The surface of the final grade is required to fall a minimum of 6 inches within the first 10 feet away from the foundation (Figure 3-8). Depending on local site conditions, it is not always possible to achieve that much fall, and the code permits alternative designs to drain the water away from the foundation. In this case, the surface water may be directed to swales or drains to ensure adequate drainage away from the structure. Impervious surfaces within 10 feet of the foundation, such as concrete driveways, sidewalks, and patios, must be sloped not less than 2 percent away from the structure (Figure 3-9). [Ref. R401.3, R404.1.6]

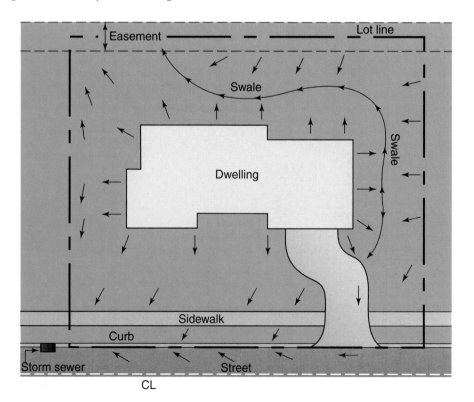

FIGURE 3-7 Drainage plan

Though the prescribed slopes as previously discussed are concerned with the first 10 feet away from the structure, the IRC also has requirements for drainage to an approved location such as a storm drain, storm sewer inlet, or the street gutter that leads to a storm drain. The drainage design must consider the entire lot for any impediments to drainage during heavy rains. **[Ref. R401.3, R403.1.7.3]**

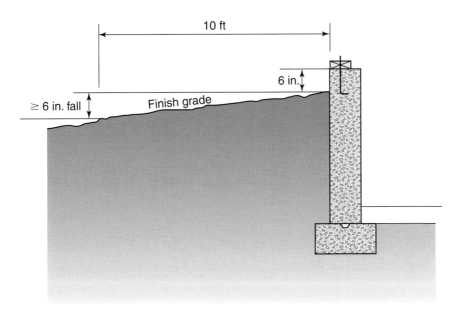

FIGURE 3-8 Grade sloped 6 inches in 10 feet to provide surface drainage away from foundation

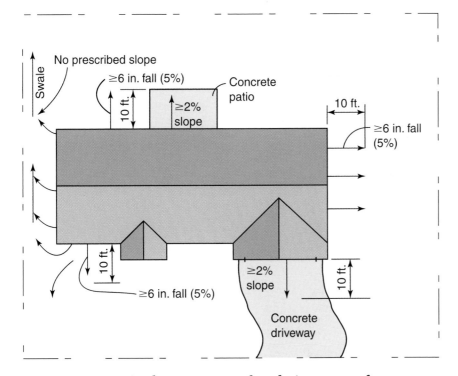

FIGURE 3-9 Grade to ensure surface drainage away from structure

FLOOD HAZARD AREAS

Site development in designated flood hazard areas must meet special requirements for flood resistant construction to minimize damage during a flood event. State or local floodplain management ordinances often supersede the IRC provisions for flood resistant construction. Requirements are similar, however, and are designed to satisfy the minimum standards set forth in the National Flood Insurance Program (NFIP), managed by the Federal Emergency Management Agency (FEMA) (Figure 3-10).

FIGURE 3-10 Flooded dwellings

PART III

Structural

Structural
Design Criteria

The *International Residential Code* (IRC) establishes minimum structural design criteria necessary to accommodate normal loads placed on a building and to resist the forces of natural hazards such as wind, snow, earthquake, and flood. In most cases, the prescriptive provisions of the code incorporate these criteria and offer a means of conventional construction without the need for an engineered design or complex calculations (Figure 4-1). To correctly apply the values of the tables and the prescriptive methods of construction, it is necessary to understand the structural design criteria based on geographic location and climate. In addition to structural considerations, the code requires the building be protected from other environmental hazards such as ice dams and termites. When adopting the IRC, the jurisdiction must provide the values for IRC Table R301.2(1), "Climatic and Geographic Design Criteria." These criteria are:

- Ground snow load
- Wind design
- Seismic design category (for earthquake)
- Weathering index

- Frost line depth
- Termite damage
- Winter design temperature
- Ice barrier underlayment requirements
- Flood hazards
- Air freezing index
- Mean annual temperature [**Ref. R301.1, R301.2**]

FIGURE 4-1 Concrete, wood, and steel structural elements

PRESCRIPTIVE AND PERFORMANCE

The intent of the code is to provide comprehensive but easy to use minimum standards for the conventional construction of residential buildings and at the same time provide the greatest design flexibility in recognizing other methods and materials of construction. With this in mind, the IRC contains both prescriptive and performance requirements. *Prescriptive* means a set of rules the builder may follow to ensure that the building complies with the code. *Performance* means an expectation that the building system will function in a certain way to meet the minimum requirements of the code. In terms of the structural requirements, performance is typically achieved through engineering.

When using the conventional construction provisions, an engineered design is necessary for only those structural elements that exceed the limits of or are otherwise not included in the prescriptive provisions of the code. For example, the sizing of wide flange steel beams commonly used in dwelling construction is outside the scope of the IRC conventional framing systems and must be designed in accordance with accepted engineering practice. This does not prevent the designer and builder from using the prescriptive methods for the rest of the building. In other words, the IRC permits partial or complete engineering of the structure and offers the prescriptive methods as an option, but they are not mandatory. The code imposes seismic, wind, and snow loading limitations on the use of the prescriptive framing methods, as will be discussed in later sections of this chapter. [**Ref. R301.1.2, R301.1.3**]

As an alternative to the wood framing provisions of the code, the IRC permits construction to comply with the *Wood Frame Construction Manual* (WFCM), published by the American Forest and Paper Association (AF&PA). The WFCM offers both engineered and prescriptive design requirements for one- and two-family dwellings. The prescriptive design requirements are based on engineering analysis. While the limits of the design provisions are generally consistent with those in the IRC, the WFCM contains provisions for construction in regions with wind speeds as great as 150 mph. The prescriptive methods of the IRC limit construction to those areas with wind speeds less than 110 mph and less than 100 mph in hurricane-prone regions (Figure 4-2). [**Ref. R301.1.1**]

FIGURE 4-2 Wood framing

BASIC LOADS (LIVE AND DEAD)

Building construction must safely support all loads, meaning the forces acting on the building. *Gravity loads* refer to the weight of objects bearing down on the structure and include live loads, dead loads, and roof loads. Live loads are the variable loads related to the use of the structure, such as people and furniture. Prescriptive design presumes uniform distribution of the live loads expressed in pounds per square foot (psf) based on the use of the space (Table 4-1). Dead loads are permanent in nature and include the weights of all construction materials and fixed equipment incorporated into the building. The prescriptive tables in the IRC include the combined effects of live loads and dead loads. The total roof load is a combination of dead and live loads, except for buildings in regions where the roof snow load exceeds the roof live load. The code assumes a roof live load—the people, materials, and equipment associated with roofing activities—at not greater than 20 psf. The roof framing is required to support the roof live load or the snow load, whichever is greater. **[Ref. R301.4, R301.5, R301.6, Table R301.5]**

Live loads

Designs for bedroom areas assume a uniform floor live load of not less than 30 psf, and all other living areas of a dwelling require a minimum live load of 40 psf. Such loads are reflected in the prescriptive tables of the code. Elevated garage floors for vehicles must be designed for a

TABLE 4.1 Minimum uniformly distributed live loads

Use	Live load (psf)	Note
Rooms other than sleeping rooms	40	
Sleeping rooms	30	
Balconies and decks	40	
Stairs	40	Concentrated load of 300 lb. / 4 sq. in. of tread
Handrails and top rails of guards		Concentrated load of 200 lb. applied in any direction
Guard balusters and infill panels		Horizontally applied load of 50 lb. on an area of 1 sq. ft.
Passenger vehicle garages	50	Elevated garage floors must support a concentrated load of 2,000 lb. / 20 sq. in.
Attics without storage	10	
Attics with limited storage	20	A storage area that is at least 24 in. wide x 42 in. high with an access hatch or pull-down stair
Attics served by a fixed stair	30	
Habitable attics	30	Floor area ≥ 70 sq. ft. meeting ceiling height requirements

minimum uniform live load of 50 psf and be capable of supporting a concentrated load of 2000 pounds on any 20-square-inch area. Typically requiring an engineered design, this criterion is necessary to accommodate the concentrated load of a vehicle transferred to the relatively small area of the tires resting on any portion of the floor.

The performance requirements of stairs and railings present some challenges to the inspector. The IRC does not include prescriptive structural design requirements for these elements, though a number of conventional and traditional methods are often deemed acceptable without requiring engineering or supporting documentation. Guards and handrails must be secured to safely resist forces that could act against them. Handrails and top rails of guards are required to be constructed so that they are capable of resisting a 200-pound concentrated force from any direction. The infill components of a guard—often spindles, balusters, or intermediate rails—must be able to safely resist 50 psf applied over a 1-square-foot area. Mechanical instruments are available to measure forces applied to railing components, but inspectors more often make a subjective determination based on rigidity of the components when weight is applied against them. The same may be said for stairs. Stairs are designed for a uniform load of 40 psf, but the treads must also be able to support a concentrated load of 300 pounds acting on an area of 4 square inches. This is a conservative value to account for the entire weight of a person on the ball of the foot bearing on the tread. One-inch and 2-inch nominal thickness dimension lumber and manufactured composite stair treads are presumed to accommodate this concentrated load. The builder and inspector often judge the 40 psf live load by experience and simply walking the stair. Stiffness or bounce of the stair should be comparable to walking across a floor.

Roof live load is a variable load that generally occurs during roofing or roof maintenance operations. The basic roof live load is 20 psf but may be reduced based on a steeper roof pitch or averaging the load effects over a larger area. For other than an engineered design, these roof live load reductions are typically not a factor in construction under the IRC because the prescriptive tables assume a roof live load of 20 psf. For areas with a roof snow load greater than 20 psf, the snow load controls the roof design.

Dead loads

Average dead loads are also included in the prescriptive tables for footings, floors, walls, and roofs. For example, spread footing sizes for conventional frame construction assume average weights of construction materials being supported. Therefore, additional calculations are typically not required. The material and component weights in Tables 4-2 and 4-3 may be helpful in sizing a post or pad footing or another structural element not covered in the IRC tables.

Deflection

Allowable deflection in structural framing members such as studs, joists, and beams is a way to ensure adequate stiffness when such members are subjected to bending under code-prescribed loads. For a floor joist, this may be understood as the bounce or give in the floor system as a person walks across a room. A design for less deflection will translate to more stiffness and therefore less bounce in the floor. The IRC sets limits on the maximum allowable deflection depending on the type of member involved. The code permits greater deflection, for example, in ceiling joists and rafters than in floor joists. These are minimum code requirements, and the homebuilder or homeowner may desire more stiffness and less bounce in a floor than the code would otherwise allow. Although deflection

TABLE 4-2 Building material weights

Materials	Weight (psf)
Plywood, ¼ in.	0.8
Plywood, ½ in.	1.6
Plywood, ¾ in.	2.4
Brick, 4 in.	35
Gypsum board, ½ in.	2.1
Gypsum board, ⅝ in.	2.5
Plaster, 1 in.	8.0
Stucco, ⅞ in.	10.0
Quarry tile, ½ in.	5.8
Hardwood flooring, 25/32 in.	4.0
Built-up roofing	6.5
Shingles, asphalt	1.7–2.8
Shingles, wood	2.0–3.0
Common dimension lumber (pcf)	27–29 pcf
Concrete (pcy)	150 pcy

pcf = pounds per cubic foot; pcy = pounds per cubic yard.

TABLE 4-3 Average weights of building components

Description	Weight (psf)
Roof dead load (framing, sheathing, asphalt shingles, insulation, drywall)	10
Exterior wall (2 × 4 framing, sheathing, siding, insulation, drywall)	10
Floor (joist, sheathing, carpeting, drywall)	10
Concrete wall, 8 in. thick	100
10 in. thick	125
12 in. thick	150
Concrete block wall, 8 in. thick	60

TABLE 4-4 Allowable deflection of structural members

Structural Member	Allowable Deflection
Rafters, slope >3/12, no finished ceiling attached to rafters	L/180
Rafters, slope >3/12, finished ceiling attached to rafters	L/240
Ceiling joist	L/240
Plastered ceilings	L/360
Floors	L/360
All other structural members	L/240

Note: Wall deflection and wind load deflections are not shown.

limitations are incorporated into the prescriptive tables, it is important to understand deflection in using the appropriate table for sizing a framing member. Allowable deflection is measured by dividing the span (L) of the member by a prescribed factor, such as 360 for floor joists. Thus the allowable deflection of a floor joist with span L inches would be L/360 inches (Table 4-4). **[Ref. R301.7, Table R301.7]**

> **EXAMPLE**
> The following example is for a floor joist with a 14-foot span.
>
> $$L = 14 \text{ ft.} \times 12 \text{ in.} = 168 \text{ in.}$$
>
> Allowable deflection = 168 in. / 360 = 0.47 in. or approximately ½ inch. Note that a 14-foot span rafter with 4:12 slope and no ceiling attached has an allowable deflection of L/180, which is twice the deflection allowed for floor joists.

WIND, SNOW, SEISMIC, AND FLOOD LOADS

In addition to supporting the live and dead loads, the building must safely resist environmental load effects such as wind, snow, earthquake, and flood hazards. These forces may be vertical (up or down) or lateral (sideways) and are also referred to as loads. **[Ref. R301.1]**

FIGURE 4-3 Wind forces acting on building

Code Basics

IRC conventional framing limits on wind speed:

• Hurricane regions, <100 mph
• Elsewhere, <110 mph

WFCM conventional framing limits on wind speed:

• 150 mph •

Wind

Lateral wind pressure may be positive (pushing against the building on the windward side) or negative (suction forces on the leeward side of a building). Wind pressure can also produce upward forces referred to as *uplift*. The building resists wind forces with wall bracing, sheathing, and positive load path connections from the roof down to the foundation (Figure 4-3).

Prescriptive design under the IRC is generally limited to those geographic regions with wind speeds less than 110 mph as defined in the IRC wind maps. The code is more restrictive in hurricane-prone regions, where the prescriptive methods are limited to wind speeds less than 100 mph. Where the basic wind speeds equal or exceed these limits, the code references several alternate methods for design. Conventional wood frame construction may comply with the WFCM or the ICC *Standard for Residential Construction in High Wind Regions* (ICC-600) (Figure 4-4). Otherwise, the code requires design in accordance with the engineering provisions of the *International Building Code* (IBC). **[Ref. R301.2.1.1]**

Exposure category

In addition to the basic wind speeds for a geographic area, ground surface irregularities affect the wind forces placed on a building. Forested terrain or groups of buildings in close proximity may shield a building from wind. Flat, open terrain, on the other hand, exposes a building to the full effects of wind.

The IRC classifies wind exposure into four categories, A to D. Basically, exposure A is limited to large

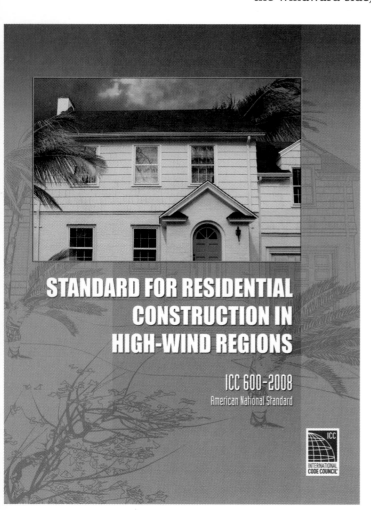

STANDARD FOR RESIDENTIAL CONSTRUCTION IN HIGH-WIND REGIONS

ICC 600-2008
American National Standard

FIGURE 4-4 ICC 600

FIGURE 4-5 Wind exposure B

FIGURE 4-6 Wind exposure C

cities with a majority of high-rise buildings, though the IBC and ASCE-7 no longer recognize exposure A for an engineered design. Exposure B, the default and most common application, affords some wind protection with trees and buildings characteristic of urban and suburban settings. Exposure C is basically open terrain with scattered obstructions and shorelines in hurricane-prone regions. Exposure D applies to buildings adjacent to large bodies of water, such as the Great Lakes and the western coastal areas, and excludes shorelines in hurricane-prone regions.

Exposure categories are important design criteria for engineering of buildings or portions of buildings resisting the effects of wind, and such criteria should appear on engineering submittal documents. Many of the prescriptive methods of wood frame construction in the IRC are generally deemed in compliance without consideration of the wind exposure category. However, wind exposure category is considered when applying the provisions for wall sheathing, wood wall bracing, and exterior wall and roof coverings. Siding, roofing, windows, skylights, exterior doors, and overhead doors must be listed and installed to resist wind loads based on roof height and exposure factors according to IRC tables R301.2(2) and R301.2(3) (Figures 4-5 and 4-6). **[Ref. R301.2.1.4]**

Hurricanes
Hurricane-prone regions are the coastal areas of the Atlantic Ocean and Gulf of Mexico where the basic wind speed is greater than 90 mph. The IRC intends to reasonably limit damage from the destructive forces of hurricanes, even though hurricane winds may significantly exceed basic design wind speeds (Figures 4-7 and 4-8). In hurricane-prone regions, when basic wind speeds are 100 mph or greater, the code requires a design complying with the WFCM, ICC-600, or the IBC. Additional protection is required for windows in windborne debris regions, those hurricane areas within one mile of the coast, or with wind speeds of 120 mph or greater (Figure 4-9). **[Ref. R301.2.1.2]**

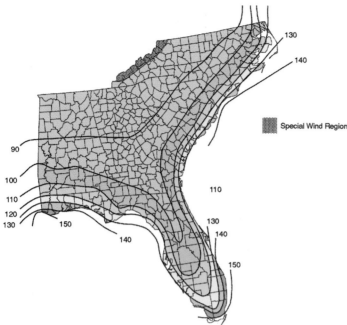

Special Wind Region

FIGURE 4-7 Example of design wind speed map

FIGURE 4-8 Hurricane

Tornadoes

The IRC does not specifically address tornadoes, whose winds may exceed 250 mph, far in excess of the basic design wind speeds. The intent of the code is to provide minimum standards of construction, including those to resist the effects of natural hazards. Buildings constructed in conformance with the code have demonstrated effectiveness in limiting damage due to high winds. The forces and effects of tornadoes are variable and unpredictable. It is not anticipated that a building of wood frame construction in the direct path of a high-intensity tornado will escape without significant damage (Figure 4-10).

Storm shelters

Storm shelters, sometimes called safe rooms, though not required by the code, offer added protection from the destructive forces of high winds, hurricanes, and tornadoes. The design and construction of storm shelters, either as detached structures or safe rooms within a dwelling, must conform to the requirements of ICC/NSSA-500, *Standard on the Design and Construction of Storm Shelters* (Figure 4-11). The International Code Council (ICC) and the National Storm Shelter Association (NSSA) jointly developed this new consensus standard. Primarily designed for life safety considerations, storm shelters protect occupants from serious injury due to high wind and flying debris. The shelters are designed to withstand impact from windborne projectiles, referred to as *missiles,* such as 2 × 4s or other construction or natural material debris that are common to high-wind events. The outer shell of above-ground shelters may be concrete, steel, composite, or other materials that have been tested to the prescribed large missile tests. **[Ref. R323]**

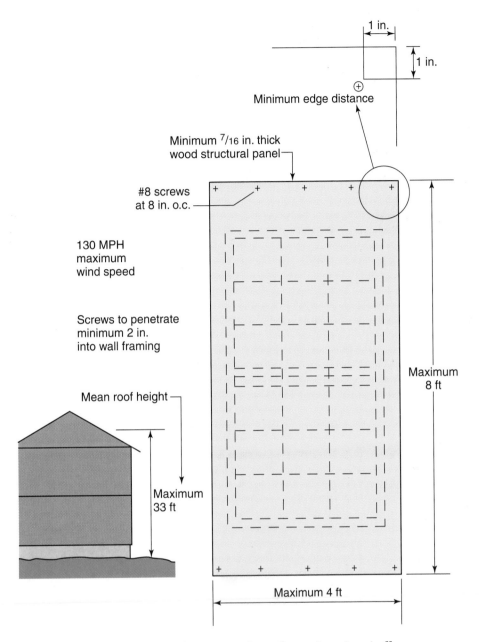

Minimum edge distance

Minimum $^7/_{16}$ in. thick wood structural panel

#8 screws at 8 in. o.c.

130 MPH maximum wind speed

Screws to penetrate minimum 2 in. into wall framing

Mean roof height

Maximum 33 ft

Maximum 8 ft

Maximum 4 ft

1 in.

1 in.

FIGURE 4-9 Prescriptive protection of openings in windborne debris regions

FIGURE 4-10 Tornado damage *(Courtesy of iStock)*

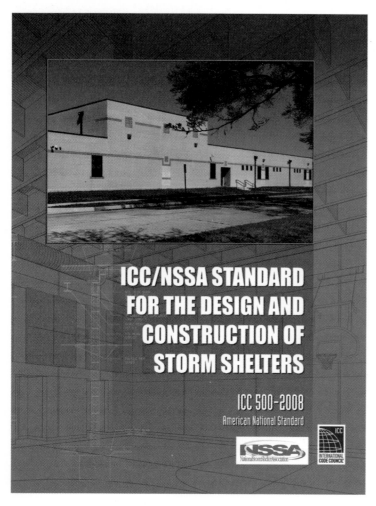

FIGURE 4-11 ICC 500

Snow

The IRC limits the use of the prescriptive provisions to buildings in areas with a ground snow load of 70 psf or less. In areas with higher snow loads, engineering is required for portions or elements supporting a snow load (Figure 4-12). The prescriptive rafter tables reflect ground snow loads of 30, 50, and 70 psf. Engineered components such as roof trusses are designed for the greater of roof live load or roof snow load. Ground snow load is the basis for determining the roof snow load design criteria for engineered components.

The map in IRC Figure R301.2(5) specifies the ground snow load for geographic regions. In some regions, indicated by "CS" on the map, and at elevations exceeding the limits shown, the ground snow load must be determined by site-specific case studies. The state or municipal authority may determine the local snow load based upon historical evidence without referencing the map. **[Ref. R301.2.3]**

Earthquake

The design and construction of buildings must resist the load effects caused by ground motion from earthquakes and limit the resulting damage. As indicated in the seismic map in the code, the IRC assigns a seismic design category (SDC) to building sites relative to the anticipated intensity and frequency of earthquakes (Figure 4-13). All buildings constructed under the prescriptive

You Should Know

Seismic design category (SDC) is a classification assigned to a structure based on its seismic group and the severity of the design earthquake ground motion at the site. ●

FIGURE 4-12 Roof snow load *(Courtesy of iStock)*

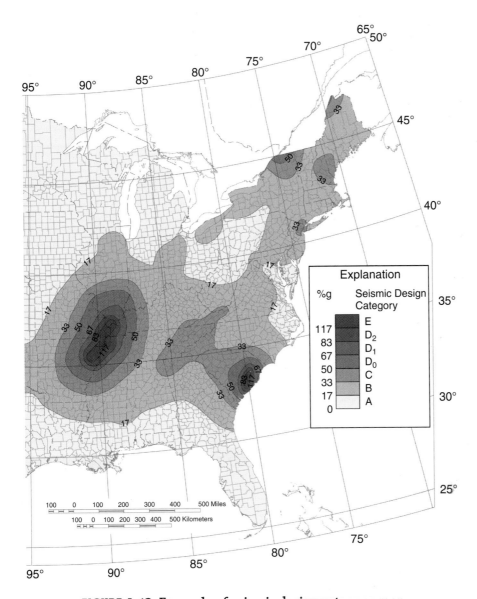

FIGURE 4-13 Example of seismic design category map

provisions of the IRC are judged to perform satisfactorily in SDCs "A" and "B." Though the conventional methods may be used in SDCs "A" to "D," specific seismic requirements apply to buildings sited in SDCs "C," "D_0," "D_1," and "D_2." However, one- and two-family dwellings are exempt from these additional requirements in SDC "C." Buildings in SDC "E" require an engineered design in accordance with the IBC. [**Ref. R301.2.2**]

Earthquake forces distort the shape of buildings as the foundation moves with the shaking ground motion and the upper portions of the building resist this movement, producing lateral forces that may cause severe damage to the structure (Figure 4-14). This resisting force is referred to as *inertia,* the tendency of an object at rest to remain at rest. Because inertia forces are directly proportional to the weights of the materials in the upper portions of the building, with heavier materials resulting in higher seismic forces and increased damage, the IRC seismic provisions impose limits on the dead load of the various roof, ceiling, wall,

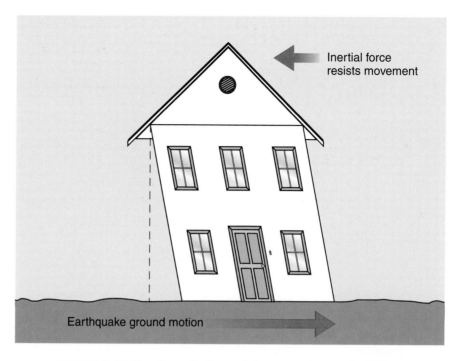

FIGURE 4-14 Earthquake lateral forces acting on building

and floor elements and may also limit the number of stories. In addition to the lateral, or side-to-side, shaking forces of earthquakes, seismic events may also produce upward forces or change the stiffness characteristics of the soil, resulting in foundation settlement.

Regular buildings have uniform distribution of forces and more predictable response characteristics when subjected to earthquake ground motion. Irregularly shaped buildings have force concentrations and are generally less effective in resisting earthquake load effects (Figure 4-15). Consequently, except for one- and two-family dwellings in SDC "C," the IRC requires engineering for portions of buildings considered irregular in SDCs "C," "D_0," "D_1," and "D_2." Among the factors determining irregular buildings are offsets in braced wall lines, arrangement of openings, cantilevers, and the use of dissimilar materials in braced wall lines. [Ref. R301.2.2.2.5]

Vertical offset in floors—floor joists are not lapped or tied together

Offset in braced wall line

FIGURE 4-15 Irregular building

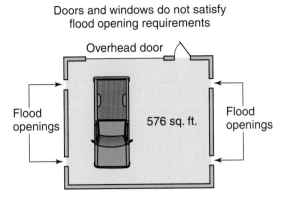

Doors and windows do not satisfy
flood opening requirements

Overhead door

Flood openings   576 sq. ft.   Flood openings

Garage below design flood elevation
used for parking and storage only

Flood opening requirements:

- Two openings required
  on different sides

- Total 576 sq. in. net area
  (1 sq. in. per sq. ft. of
  enclosed area)

- 576/4 = 144 sq. in. net area
  per opening

- The bottom of each
  opening no more than 12 in.
  above grade

FIGURE 4-16 Flood openings

Floods

The IRC requires flood resistant construction for buildings located in flood hazard areas to minimize damage. Primary protection is achieved through elevation of the lowest floor of the building above the design flood elevation. Basements and spaces used only for storage or vehicle parking are permitted below the design flood elevation. Such enclosed areas require flood openings to allow flood waters to flow through the space and equalize hydrostatic pressure on both sides of the enclosing walls (Figure 4-16). The structural elements of these spaces, as well as building foundations, must effectively resist flood loads in addition to other applicable loads. Flood loads, velocities, and hazards associated with floating debris are increased in floodways, the main channel boundaries of a river. Buildings located in a floodway must be designed and constructed in accordance with ASCE 24 to adequately resist these loads. **[Ref. R301.2.4, R322.2.2, R322.2.3]**

The *International Residential Code* (IRC) contains prescriptive foundation designs to safely support building loads and transmit those loads to the soil. This chapter will focus primarily on the provisions for conventional concrete footings and concrete and masonry foundation walls (Figure 5-1). Soil characteristics and bearing capacities affecting foundation design are discussed in Chapter 3.

FIGURE 5-1 Concrete foundation for single-family dwelling

MATERIALS

The two most common materials for foundation construction are concrete and concrete block, the latter more precisely described as *concrete masonry units* (CMUs). The code does not intend to limit the use of different materials, however. In addition to prescriptive designs for other foundation systems incorporating wood, precast concrete, or *insulating concrete forms* (ICFs), the IRC permits engineered or alternative designs. **[Ref. R402]**

Concrete

Concrete continues to gain strength after the initial set through a chemical curing process. The compressive strength of concrete is related to the proportions of portland cement, sand, gravel, and water in the mix and is expressed in pounds per square inch (psi) after 28 days' curing time. The code requires concrete to have a minimum 28-day compressive strength of 2500 psi for most applications (Table 5-1). Higher-strength concrete, often including entrained air, is specified in geographic areas

TABLE 5-1 Minimum specified compressive strength of concrete

Type or Location of Concrete Construction	Minimum Specified Compressive Strength at 28 Days (psi)		
	Weathering Potential		
	Negligible	Moderate	Severe
Basement walls, foundations, and other concrete not exposed to the weather	2500	2500	2500[a]
Basement slabs and interior slabs on grade, except garage floor slabs	2500	2500	2500[a]
Basement walls, foundation walls, exterior walls, and other vertical concrete work exposed to the weather	2500	3000[b]	3000[b]
Porches, carport slabs, and steps exposed to the weather, and garage floor slabs	2500	3000[b]	3500[b]

a. Concrete subject to freezing and thawing during construction shall be air entrained.
b. Concrete shall be air entrained.

FIGURE 5-2 Concrete placement

subject to moderate or severe weathering potential when the concrete is exposed to the weather or is for a garage floor slab (Figure 5-2). [Ref. Table R402.2]

FOOTINGS

In order to properly support the loads of a building, footing design must address not only the size of the footing, but factors such as the condition and characteristics of the soil, footing depth, slope, and reinforcing. [Ref. R403]

Depth, bearing, and slope

For other than engineered soil conditions, footings must bear on undisturbed ground and extend below the frost depth to provide a stable foundation. In addition, exterior footings require excavation to at least 12 inches below the undisturbed soil. Vegetation, wood, debris, loose or frozen soil, and any other detrimental materials are removed prior to placing concrete (Figure 5-3). [Ref. R403.1.4]

Placing footings below the frost depth protects foundations from the expansive effects of freezing and thawing soil. Otherwise, frost heave can exert stresses sufficient to cause significant damage to a foundation. The code offers exceptions to the frost depth requirements where damaging effects of frost heave are negligible. Accessory buildings with an eave height of 10 feet or less and limited in area—600 square feet for light frame construction and 400 square feet for other construction—and decks not attached to a dwelling do not require frost protection. The code also permits frost-protected shallow foundations utilizing rigid polystyrene insulation.

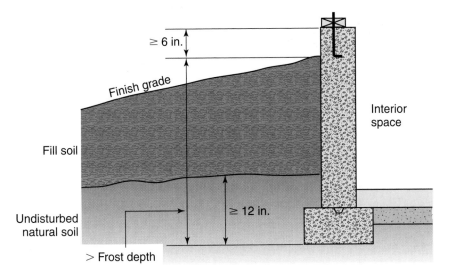

FIGURE 5-3 Depth of footings

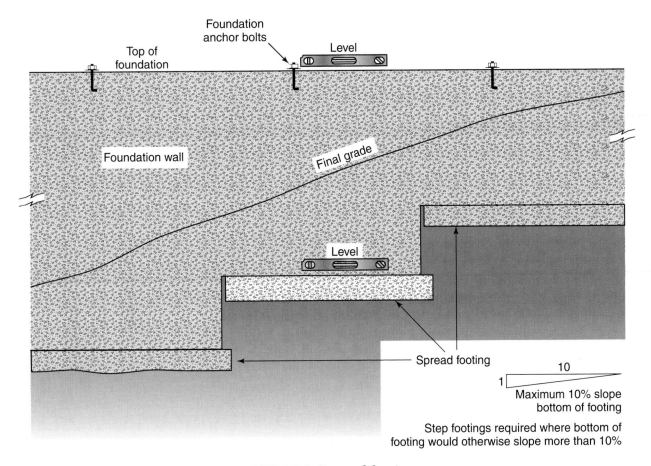

FIGURE 5-4 Stepped footing

To prevent sliding and to adequately transfer loads to the soil, the code limits the slope of the bottom of footings to a maximum 1 unit vertical in 10 units horizontal (10 percent slope). Transitions that would result in greater slopes must be achieved through stepping of the footings (Figure 5-4). **[Ref. R403.1.5]**

TABLE 5-2 Minimum width of concrete footings (in.)

	Load-Bearing Value of Soil (psf)			
	1500	**2000**	**3000**	**≥4000**
Conventional Light-Frame Construction				
1-story	12	12	12	12
2-story	15	12	12	12
3-story	23	17	12	12
4-in. Brick Veneer over Light-Frame or 8-in. Hollow Concrete Masonry				
1-story	12	12	12	12
2-story	21	16	12	12
3-story	32	24	16	12
8-in. Solid or Fully Grouted Masonry				
1-story	16	12	12	12
2-story	29	21	14	12
3-story	42	32	21	16

Sizing concrete footings

Soil-bearing capacity and the average gravity loads (dead, live, and snow) determine footing size. As the load-bearing capacity of the soil decreases, footing size increases to distribute the load to a greater area. For example, soil with a bearing capacity of 1500 psf will require a footing with twice the bearing area as soil with a bearing capacity of 3000 psf when supporting the same total gravity load. (Such an increase is not readily apparent in some of the values in Table 5-2 because the IRC does not permit footings with a width less than 12 inches, and the table averages typical gravity loads in determining the prescriptive widths of footings). Footings must have sufficient bearing area to prevent differential settlement, which can cause structural and performance problems. The IRC prescribes the width of continuous footings based on the number of stories supported, the method of construction, and the load-bearing value of the soil (Figure 5-5). The minimum thickness of footings is 6 inches (Table 5-2 and Table 3-2 in Chapter 3). [**Ref. R403.1.1 and Table R403.1**]

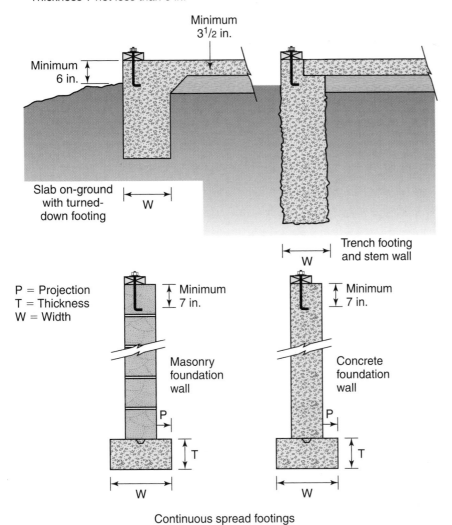

Width W per Table 5-2
Projection P at least 2 in. and not greater than T
Thickness T not less than 6 in.

Minimum
$3^{1}/_{2}$ in.

Minimum
6 in.

Slab on-ground
with turned-
down footing

W

Trench footing
and stem wall

W

P = Projection
T = Thickness
W = Width

Minimum
7 in.

Masonry
foundation
wall

P

T

W

Minimum
7 in.

Concrete
foundation
wall

P

T

W

Continuous spread footings

FIGURE 5-5 Types of continuous footings

EXAMPLE

Size of a continuous footing (width and depth)

Determine minimum width (W), projection (P), and thickness (T) of a continuous spread footing (Figure 5-5) for a two-story dwelling (Figure 5-6) assuming 1500 psf soil bearing and using the prescriptive values of Table 5-2. Projection (P) must be at least 2 inches and cannot exceed thickness (T). The minimum footing thickness (T) is 6 inches. The solution is shown in Figure 5-7.

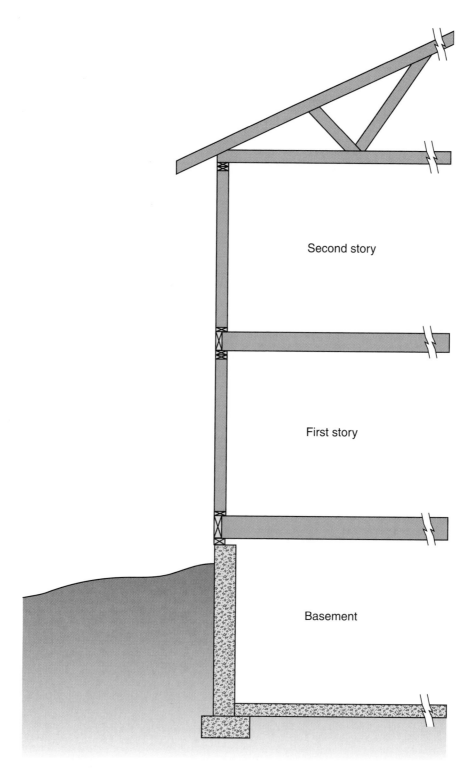

FIGURE 5-6 Cross section of two-story dwelling for determining footing size in Figure 5-7

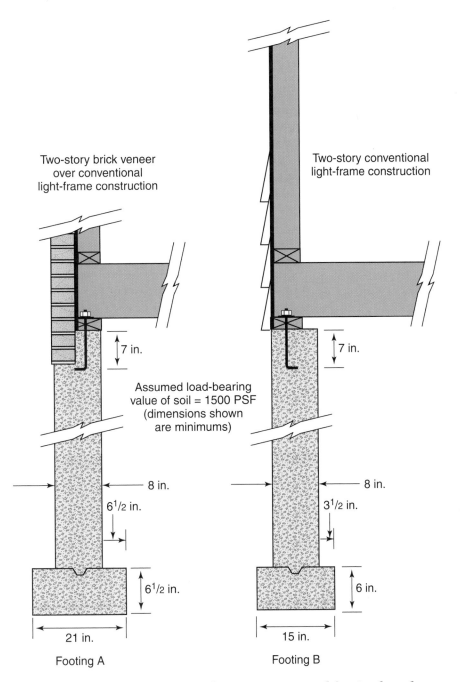

Two-story brick veneer over conventional light-frame construction

Two-story conventional light-frame construction

7 in.

7 in.

Assumed load-bearing value of soil = 1500 PSF (dimensions shown are minimums)

8 in.

6¹/₂ in.

6¹/₂ in.

21 in.

Footing A

8 in.

3¹/₂ in.

6 in.

15 in.

Footing B

FIGURE 5-7 Minimum size of continuous spread footing based on Table 5-2 and Figure 5-6

EXAMPLE

Size of isolated footing for a column load

The soil type and the total tributary load being supported determine pier and column footing size. See Figure 5-8 for a simple example of tributary floor load transferred to a column. Assuming a total floor load of 50 psf (40 psf live load plus 10 psf dead load), a tributary area of 120 square feet will result in a column load of 6000 pounds ($120 \times 50 = 6000$). A post carrying a 6000-pound load will require a footing area of 4 square feet (2 feet by 2 feet) for a soil-bearing capacity of 1500 psf ($6000/1500 = 4$). Minimum footing dimensions are shown in Figure 5-9.

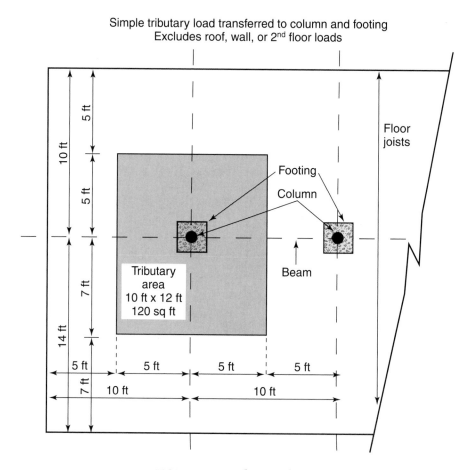

FIGURE 5-8 Tributary load

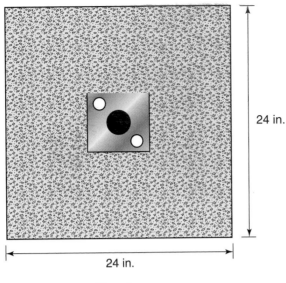

24 in.

24 in.

Plan view

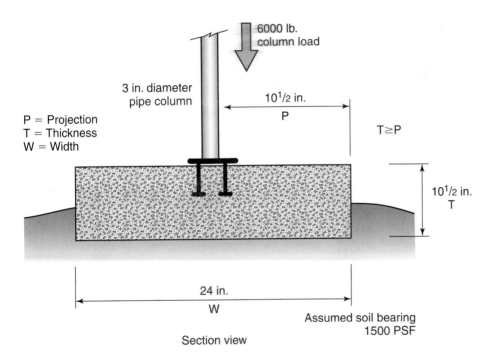

6000 lb.
column load

3 in. diameter
pipe column

P = Projection
T = Thickness
W = Width

$10^1/_2$ in.
P

T≥P

$10^1/_2$ in.
T

24 in.
W

Assumed soil bearing
1500 PSF

Section view

FIGURE 5-9 Minimum dimensions of an isolated column footing based on tributary load and assumed soil-bearing capacity

Seismic reinforcing for footings

The IRC permits footings without reinforcement—referred to as *plain concrete*—in seismic design categories (SDCs) "A," "B," and "C." There are minimum reinforcement requirements for buildings located in SDCs "D_0," "D_1," and "D_2." However, longitudinal reinforcement is not required in footings of one- and two-family dwellings (Figures 5-10 to 5-12). [Ref. R403.1.3]

FIGURE 5-10 Concrete footings

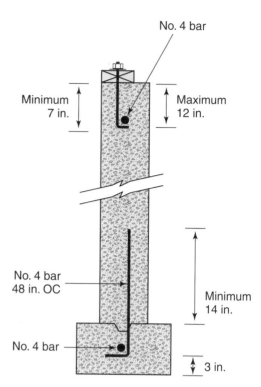

FIGURE 5-11 Footing and stem wall reinforcing in SDCs "D_0," "D_1," and "D_2"

Foundation anchorage

Anchorage to the foundation is a critical part of the load path to resist lateral and uplift forces acting on the framing system of the building. The IRC prescribes anchor bolt criteria for connecting the sill plate to the foundation. Other methods, such as foundation straps, may be used if installed according to the manufacturer's instructions and in a way to provide equivalent anchorage. Such alternatives typically require closer spacing than for embedded anchor bolts (Figure 5-13). **[Ref. R403.1.6]**

To resist the increased forces of earthquake ground motion, additional anchorage requirements apply to buildings located in SDCs "D_0," "D_1," and "D_2" and to townhouses in SDC "C." In this case, anchor bolts in braced wall lines (described in Chapter 6) require 3-inch by 3-inch plate washers approximately ¼ inch thick to increase the holding capacity against the sill plate or bottom plate (Figure 5-14). Bolt spacing is also reduced to 4 feet for anchorage of three-story buildings. One- and two-family dwellings are exempt from specific seismic design requirements in SDC "C" (Table 5-3). **[Ref. R403.1.6.1]**

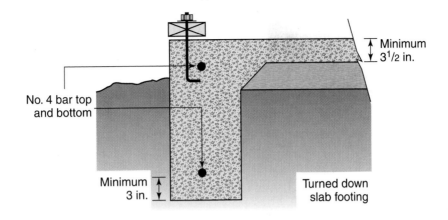

No. 4 bar top
and bottom

Minimum
3½ in.

Minimum
3 in.

Turned down
slab footing

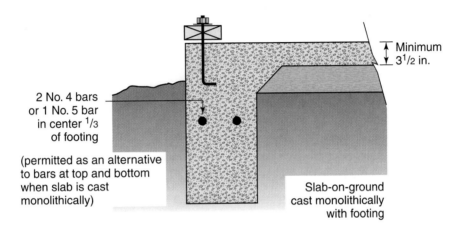

2 No. 4 bars
or 1 No. 5 bar
in center ⅓
of footing

(permitted as an alternative
to bars at top and bottom
when slab is cast
monolithically)

Minimum
3½ in.

Slab-on-ground
cast monolithically
with footing

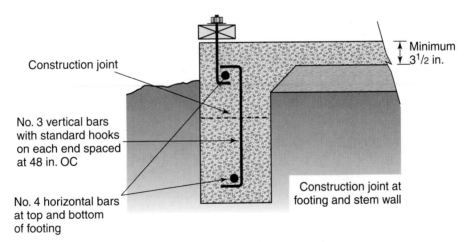

Construction joint

No. 3 vertical bars
with standard hooks
on each end spaced
at 48 in. OC

No. 4 horizontal bars
at top and bottom
of footing

Minimum
3½ in.

Construction joint at
footing and stem wall

FIGURE 5-12 Reinforcing for turned down slab footings in SDCs "D_0," "D_1," and "D_2"

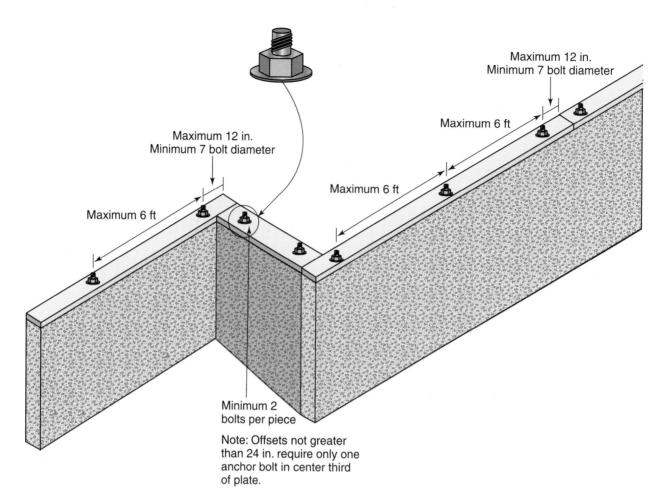

Maximum 12 in.
Minimum 7 bolt diameter

Maximum 6 ft

Maximum 12 in.
Minimum 7 bolt diameter

Maximum 6 ft

Maximum 6 ft

Minimum 2
bolts per piece

Note: Offsets not greater
than 24 in. require only one
anchor bolt in center third
of plate.

FIGURE 5-13A Wood sill plate anchorage to foundation for all buildings in SDCs "A" and "B," and dwellings in SDC "C"

FIGURE 5-13B Anchor bolts for all buildings in SDCs "A" and "B," and dwellings in SDC "C"

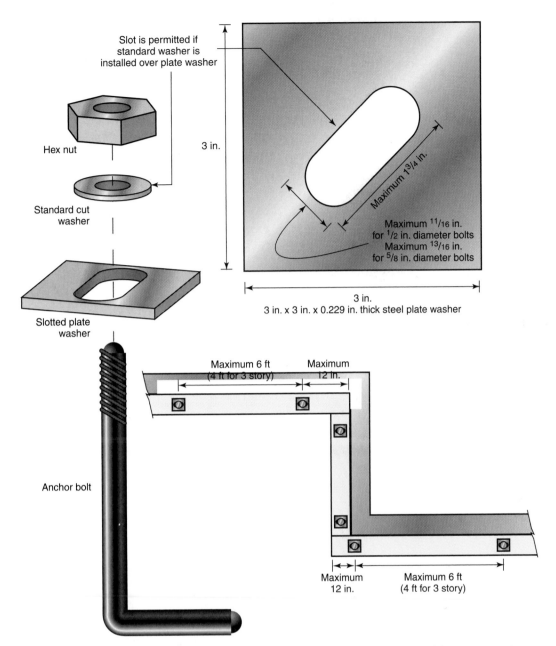

Slot is permitted if standard washer is installed over plate washer

Hex nut

Standard cut washer

Slotted plate washer

Anchor bolt

3 in.

Maximum 1³/₄ in.

Maximum ¹¹/₁₆ in. for ¹/₂ in. diameter bolts
Maximum ¹³/₁₆ in. for ⁵/₈ in. diameter bolts

3 in.
3 in. x 3 in. x 0.229 in. thick steel plate washer

Maximum 6 ft (4 ft for 3 story)   Maximum 12 in.

Maximum 12 in.   Maximum 6 ft (4 ft for 3 story)

FIGURE 5-14 Wood sill plate anchorage to foundation for all buildings in SDCs "D₀," "D₁," and "D₂" and townhouses in SDC "C"

TABLE 5-3 Foundation anchor bolts (minimum ½-inch diameter and 7-inch embedment)

Seismic Design Category	One- and Two-Family Dwellings			Seismic Design Category	Townhouses		
	Stories	Maximum Spacing	Washer		Stories	Maximum Spacing	Washer
A and B				**A and B**	1, 2, or 3	6 ft.	Standard cut
C	1, 2, or 3	6 ft.	Standard cut	**C**	1 or 2	6 ft.	3 in. × 3 in. × 0.229 in.
					3	4 ft.	
D₀, D₁, and D₂	1 or 2	6 ft.	3 in. × 3 in. × 0.229 in.	**D₀, D₁, and D₂**	1 or 2	6 ft.	
	3	4 ft.			3	4 ft.	

Note: Standard cut washers are permitted in wall lines without braced wall panels.

MASONRY AND CONCRETE FOUNDATION WALLS

The IRC includes prescriptive methods for the construction of foundation walls that are supported at their top and bottom. Top support to prevent lateral movement is typically achieved through adequate connections to the floor system. Bottom lateral support is provided by the footing in the ground or by the placement of the concrete basement slab floor directly against the foundation wall. Foundation walls that exceed the limits of the prescriptive methods in the code must be designed and constructed in accordance with the referenced standards or accepted engineering practices. **[Ref. R404.1 and R303.1.3]**

Wall height and thickness

Unlike footings, where gravity loads are the primary consideration, foundation walls must be constructed to resist lateral loads, particularly from soil pressure. Therefore, the soil type and height of the foundation determine the wall thickness and reinforcement of masonry and concrete foundation walls without consideration of the height or number of stories of the dwelling. Soil descriptions and types are given in Table 3-2 in Chapter 3. Examples of how to determine the thickness of a concrete foundation wall are given in Table 5-4 and Figures 5-15 and 5-16. **[Ref. R404.1.1 and Table R404.1.2(8)]**

Examples of how to determine the thickness of a masonry foundation wall are given in Table 5-5 and Figures 5-17 and 5-18. **[Ref. R404.1.1 and Tables R404.1.1(1) through R404.1.1(4)]**

TABLE 5-4 Minimum vertical reinforcement for concrete basement walls

Wall Height (feet)	Height of Unbalanced Backfill (feet)	Minimum Vertical Reinforcement—Bar Size and Spacing (inches)											
		Soil Classes and Lateral Soil Load (psf per foot of depth)											
		GW, GP, SW, and SP				GM, GC, SM, SM-SC, and ML				SC, ML-CL, and inorganic CL			
		30				45				60			
		Minimum nominal wall thickness (inches)											
		6	8	10	12	6	8	10	12	6	8	10	12
9	6	4 @ 34	NR	NR	NR	6 @ 48	NR	NR	NR	6 @ 36	6 @ 39	NR	NR
	7	5 @ 36	NR	NR	NR	6 @ 34	5 @ 37	NR	NR	6 @ 33	6 @ 38	5 @ 37	NR
	8	6 @ 38	5 @ 41	NR	NR	6 @ 33	6 @ 38	5 @ 37	NR	6 @ 24	6 @ 29	6 @ 39	4 @ 48
	9	6 @ 34	6 @ 46	NR	NR	6 @ 26	6 @ 30	6 @ 41	NR	6 @ 19	6 @ 23	6 @ 30	6 @ 39

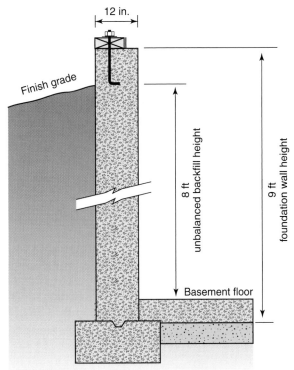

FIGURE 5-15 Wall thickness for plain concrete (no reinforcing) basement wall for all buildings in SDC A and B, and dwellings in SDC C, based on Table 5-4

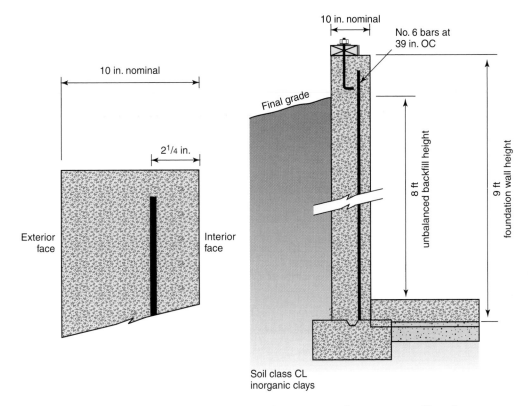

FIGURE 5-16 Wall thickness for flat concrete basement wall with reinforcing based on Table 5-4

TABLE 5-5 Masonry foundation walls

Plain Masonry Walls

| Maximum Wall Height (feet) | Maximum Unbalanced Backfill Height (feet) | Plain Masonry Minimum Nominal Wall Thickness (inches) | | |
| | | Soil Classes | | |
		GW, GP, SW, and SP Soils	GM, GC, SM, SM-SC, and ML Soils	SC, ML-CL, and Inorganic CL Soils
8	6	10	12	12 fully grouted or solid masonry
	7	12	12 fully grouted or solid masonry	Reinforcing or design required
	8	10 fully grouted or solid masonry	12 fully grouted or solid masonry	Reinforcing or design required

Eight-Inch Masonry Foundation Walls with Reinforcing Where d >5 inches

Wall Height	Unbalanced Backfill Height	Minimum Vertical Reinforcement		
		Soil Classes and Lateral Soil Load (psf per foot below grade)		
		GW, GP, SW, and SP Soils	GM, GC, SM, SM-SC, and ML Soils	SC, ML-CL, and Inorganic CL Soils
		30	45	60
8 feet 8 inches	6 feet	#4 at 48" O.C.	#5 at 48" O.C.	#6 at 48" O.C.
	7 feet	#5 at 48" O.C.	#6 at 48" O.C.	#6 at 40" O.C.
	8 feet 8 inches	#6 at 48" O.C.	#6 at 32" O.C.	#6 at 24" O.C.

Ten-Inch Masonry Foundation Walls with Reinforcing Where d >6.75 inches

Wall Height	Unbalanced Backfill Height	Minimum Vertical Reinforcement		
		Soil Classes and Lateral Soil Load (psf per foot below grade)		
		GW, GP, SW, and SP Soils	GM, GC, SM, SM-SC, and ML Soils	SC, ML-CL, and Inorganic CL Soils
		30	45	60
8 feet 8 inches	6 feet	#4 at 56" O.C.	#4 at 56" O.C.	#5 at 56" O.C.
	7 feet	#4 at 56" O.C.	#5 at 56" O.C.	#6 at 56" O.C.
	8 feet 8 inches	#5 at 56" O.C.	#6 at 48" O.C.	#6 at 32" O.C.

Twelve-Inch Masonry Foundation Walls with Reinforcing Where d >8.75 inches

Wall Height	Unbalanced Backfill Height	Minimum Vertical Reinforcement		
		Soil Classes and Lateral Soil Load (psf per foot below grade)		
		GW, GP, SW, and SP Soils	GM, GC, SM, SM-SC, and ML Soils	SC, ML-CL, and Inorganic CL Soils
		30	45	60
8 feet 8 inches	6 feet	#4 at 72" O.C.	#4 at 72" O.C.	#5 at 72" O.C.
	7 feet	#4 at 72" O.C.	#5 at 72" O.C.	#6 at 72" O.C.
	8 feet 8 inches	#5 at 72" O.C.	#7 at 72" O.C.	#6 at 48" O.C.

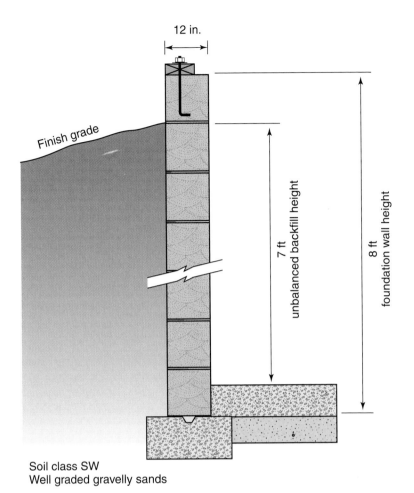

12 in.

Finish grade

7 ft
unbalanced backfill height

8 ft
foundation wall height

Soil class SW
Well graded gravelly sands

FIGURE 5-17 Wall thickness for plain masonry
(no reinforcing) foundation wall for all buildings in SDCs
"A" and "B," and dwellings in SDC "C" based on Table 5-5

Seismic requirements

Because earthquake ground motion causes significant damage to unreinforced masonry and concrete walls, the IRC places additional limitations on these foundation walls in SDCs "D_0," "D_1," and "D_2." In general, the code limits the wall height to 8 feet, with a maximum unbalanced backfill height of 4 feet and a minimum nominal wall thickness of 8 inches. This does not affect the tabular values for concrete and masonry walls with reinforcing. [Ref. R404.1.4]

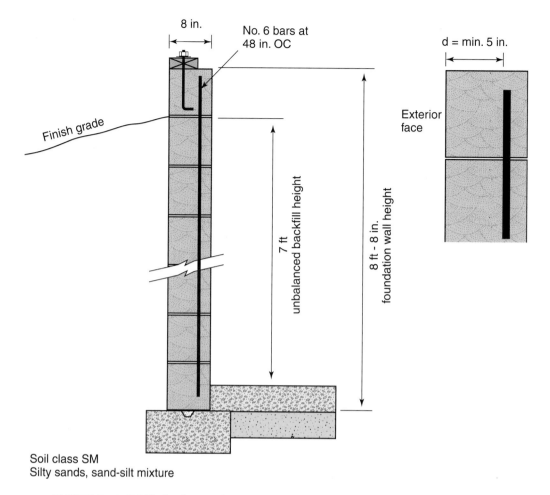

FIGURE 5-18 Wall thickness for masonry foundation wall with reinforcing based on Table 5-5

Height above finished grade

The IRC intends to protect the building from moisture intrusion that may damage portions of the structure and cause an unhealthy living environment. As part of these requirements, concrete and masonry foundation walls must extend above the finished grade a minimum of 6 inches. Masonry veneer provides somewhat better protection against moisture intrusion at ground level than do siding and other exterior wall finishes, and the elevation requirement for the top of the foundation wall is reduced accordingly to 4 inches above finished grade (Figure 5-19). **[Ref. R404.1.6]**

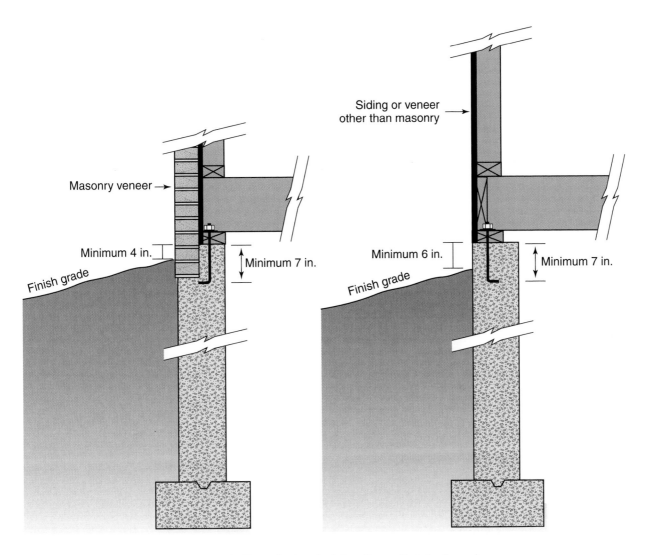

FIGURE 5-19 Foundation height above finished grade

MOISTURE PROTECTION

For foundations that retain earth and enclose spaces located below grade, the IRC requires foundation drainage and dampproofing or waterproofing to prevent moisture from penetrating into such spaces (Figure 5-20).

Foundation drainage

Except in areas with well-drained soils, the code requires perforated pipe or other approved drains to carry groundwater away from the foundation. Such drains must be installed at or below the level of the basement or crawl space floor using approved methods and the prescribed amounts of washed gravel. The drainage system is required to discharge by gravity or mechanical means to an approved location. **[Ref. R405]**

Dampproofing and waterproofing

Under normal conditions and in combination with foundation drains, dampproofing is deemed adequate to prevent moisture from migrating

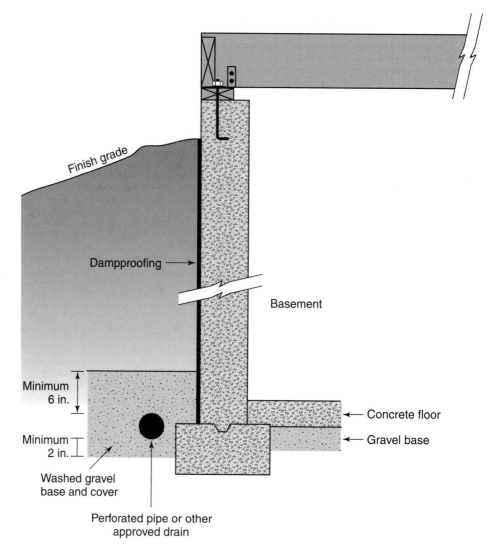

Finish grade

Dampproofing →

Basement

Minimum 6 in.

Minimum 2 in.

Washed gravel base and cover

Perforated pipe or other approved drain

→ Concrete floor

← Gravel base

FIGURE 5-20 Foundation drain and dampproofing

into the enclosed space. A bituminous-based coating or other approved dampproofing materials are applied to the exterior of the foundation from the top of the footing to the finished grade. Areas with a high water table or other known severe soil-water conditions require waterproofing. Typically consisting of flexible sealants or other impervious material and applied in thicker coatings, waterproofing provides a higher level of protection against moisture under hydrostatic pressure. **[Ref. R406]**

UNDERFLOOR SPACE

Depending on climatic conditions, significant amounts of condensation can accumulate in enclosed crawl spaces, causing decay and other damage to the structure. The code requires ventilation openings through the foundation or exterior walls in the prescribed size and location to circulate air and dissipate condensation (Figure 5-21).

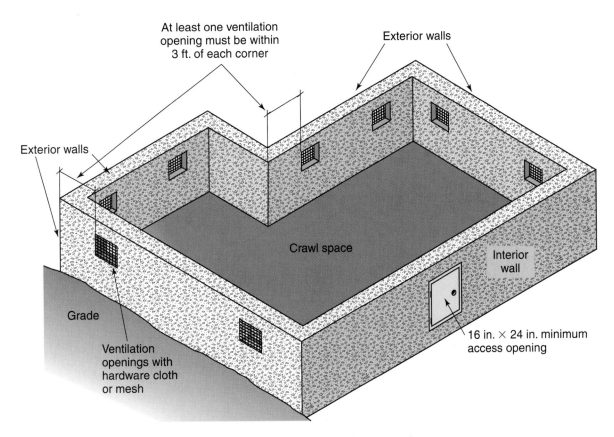

At least one ventilation
opening must be within
3 ft. of each corner

Exterior walls

Exterior walls

Crawl space

Interior
wall

Grade

16 in. × 24 in. minimum
access opening

Ventilation
openings with
hardware cloth
or mesh

FIGURE 5-21 Crawl space ventilation and access

An alternative method permits a crawl space without foundation openings when equipped with mechanical exhaust ventilation or connection to the conditioned air supply of the dwelling. In this case, the code requires insulation of the exterior walls and a vapor retarder over the ground and sealed to the enclosing foundation wall.

The IRC requires access to the underfloor spaces. Access openings through the floor must be at least 18 inches by 24 inches but may be reduced to not less than 16 inches by 24 inches when access is through a perimeter wall. [**Ref. R408**]

Framing

The repetitive system of wood or cold-formed steel framing members forming the structural elements of floor, wall, and roof construction is referred to as *light-frame construction* (Figure 6-1). The framing system, with its connections and bracing, must resist the code-prescribed vertical and lateral forces that act on the building. These loads must be adequately transferred through the framing system by a complete load path to the foundation. The *International Residential Code* (IRC) prescribes specific framing requirements that when followed preclude the need for an engineered design. This chapter will focus on the prescriptive provisions for conventional wood light-frame construction.

FIGURE 6-1 Wood framing

GRADE MARKS

Load-bearing dimension lumber for framing members and wood structural panels must be identified by a grade mark (Figure 6-2). Sawn lumber grade marks indicate the wood species, grade, moisture content, grading agency, and lumber mill identification. Species and grade determine, in part, the strength and stiffness properties that establish the maximum permissible spans for wood beams, joists, and rafters. Wood structural panel grade marks include the maximum span ratings for roof and floor applications to meet minimum performance requirements. **[Ref. R502.1, R602.1, and R802.1]**

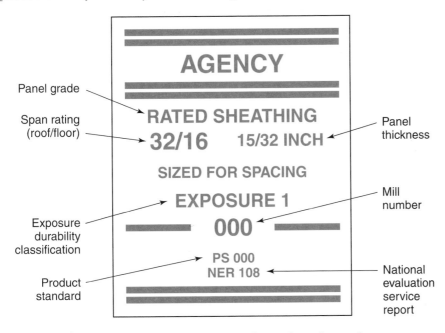

FIGURE 6-2 Wood structural panel grade mark

FIGURE 6-3 Engineered wood products

ENGINEERED WOOD PRODUCTS

Engineered wood products include plate-connected open web trusses, I-joists, glued laminated lumber, laminated veneer lumber, and other structural composite lumber. The IRC permits the use of engineered components in otherwise prescriptive conventional framing systems. These engineered components must be designed in accordance with approved engineering practice and the applicable referenced standards. Installation of engineered wood products must conform to the manufacturer's installation instructions (Figure 6-3).

TRUSSES

In addition to structural design criteria, truss design drawings include manufacturing and installation specifications for each truss (Figures 6-4 and 6-5). The IRC requires the manufacturer or contractor to submit the truss design drawings to the building official for review and approval prior to truss installation. Because they contain permanent bracing details, nailing specifications for bracing and multiple-member trusses, and minimum bearing and other important installation information, the truss design drawings must also be delivered to the jobsite with the trusses. The floor or roof sheathing and gypsum board ceiling materials typically provide adequate lateral support and thus satisfy the permanent bracing requirements for the top and bottom chords of the truss. In order to adequately resist the design loads and resist buckling, truss web members in compression often require additional lateral support that is provided by the specified permanent

FIGURE 6-4 Open web floor trusses

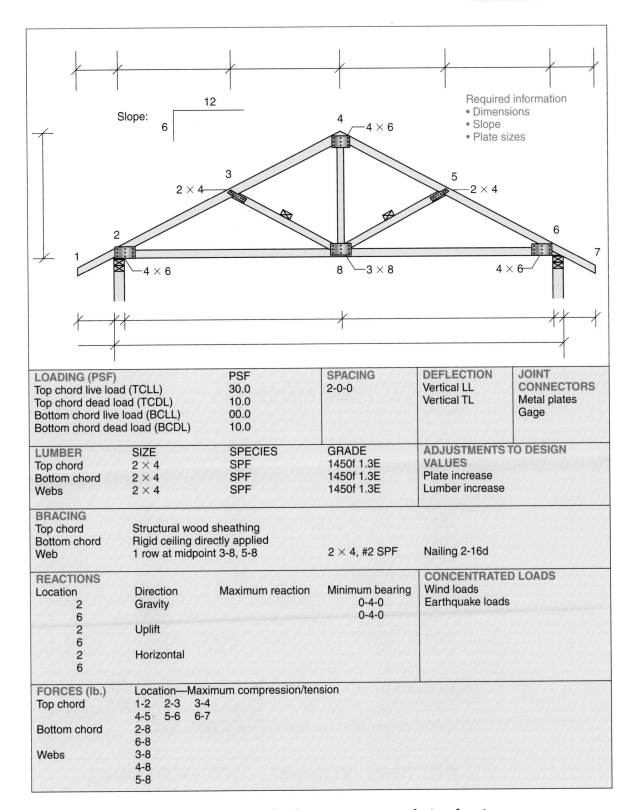

FIGURE 6-5 Required information on truss design drawing

bracing. Alterations to trusses are not allowed without the approval of a registered design professional. Approved connectors rated to resist the uplift forces indicated on the design drawings, but not less than 175 pounds, are required for the roof truss-to-wall connection. **[Ref. R502.11 and R802.10]**

FIGURE 6-6 Required label for preservative-treated wood

WOOD TREATMENT

Portions of wood construction in locations subject to decay require naturally durable wood or wood treated with preservatives. The heartwoods of decay-resistant redwood, cedar, black locust, and black walnut are considered naturally durable. An approved quality mark or label is required on preservative–treated lumber and plywood, indicating that the products meet the standards of the American Wood Protection Association (AWPA) (Figure 6-6). Preservative-treated wood suitable for ground contact is required for structural supports that are in contact with the ground, embedded in concrete in contact with the ground, or embedded in concrete exposed to the weather. Naturally durable wood is not permitted in these ground contact locations. Some chemicals used in the preservative treatment process are corrosive to steel. To resist corrosion and maintain structural load capacity, the code generally requires fasteners and connectors used in preservative and fire–retardant treated wood to be hot-dipped, zinc-coated galvanized steel, stainless steel, silicon bronze, copper, or material recommended by the manufacturer (Figures 6-7 and 6-8). [Ref. R317]

CUTTING, BORING, AND NOTCHING

In order to maintain the structural strength and integrity of wood framing, the code limits the amount and location of bored holes and notches in dimension lumber, as shown in Figures 6-9 through 6-12. The IRC generally prohibits boring, cutting, or notching of trusses and other engineered wood products except as specifically permitted by the manufacturer. Otherwise, a registered design professional must consider any such alterations in the design of the engineered component. [Ref. R502.8, R602.6, and R802.7]

Deck

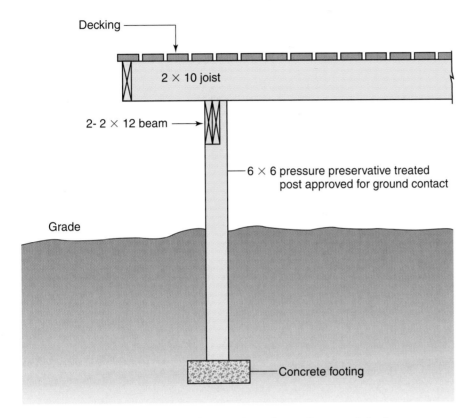

Naturally durable or preservative treated
wood required for beam, joist, and decking

Decking

2 × 10 joist

2- 2 × 12 beam

6 × 6 pressure preservative treated
post approved for ground contact

Grade

Concrete footing

FIGURE 6-7 Protection against decay for wood decks

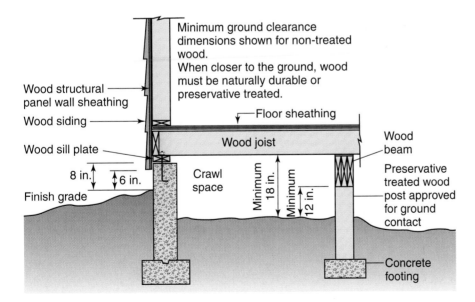

Minimum ground clearance
dimensions shown for non-treated
wood.
When closer to the ground, wood
must be naturally durable or
preservative treated.

Wood structural
panel wall sheathing

Wood siding

Floor sheathing

Wood joist

Wood
beam

Wood sill plate

8 in.

6 in.

Crawl
space

Minimum 18 in.

Minimum 12 in.

Preservative
treated wood
post approved
for ground
contact

Finish grade

Concrete
footing

FIGURE 6-8 Clearance above ground for protection against decay

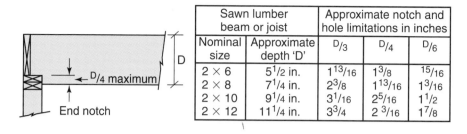

Sawn lumber beam or joist		Approximate notch and hole limitations in inches		
Nominal size	Approximate depth 'D'	$D/3$	$D/4$	$D/6$
2 × 6	$5\frac{1}{2}$ in.	$1\frac{13}{16}$	$1\frac{3}{8}$	$\frac{15}{16}$
2 × 8	$7\frac{1}{4}$ in.	$2\frac{3}{8}$	$1\frac{13}{16}$	$1\frac{3}{16}$
2 × 10	$9\frac{1}{4}$ in.	$3\frac{1}{16}$	$2\frac{5}{16}$	$1\frac{1}{2}$
2 × 12	$11\frac{1}{4}$ in.	$3\frac{3}{4}$	$2\frac{3}{16}$	$1\frac{7}{8}$

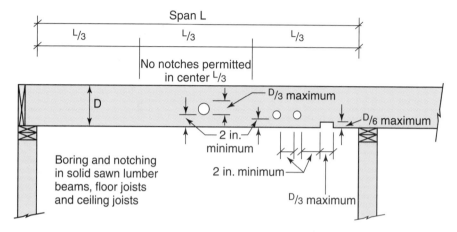

FIGURE 6-9 Boring and notching in solid sawn beams, floor joists, and ceiling joists

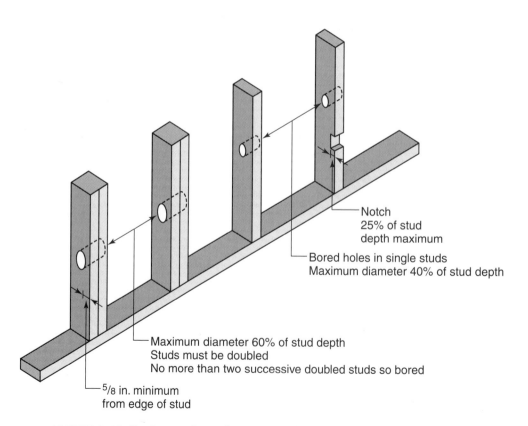

FIGURE 6-10 Boring and notching of studs in exterior wall or bearing interior wall

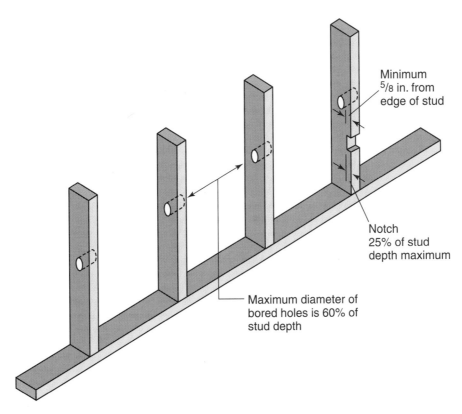

FIGURE 6-11 Boring and notching of studs in nonbearing interior wall

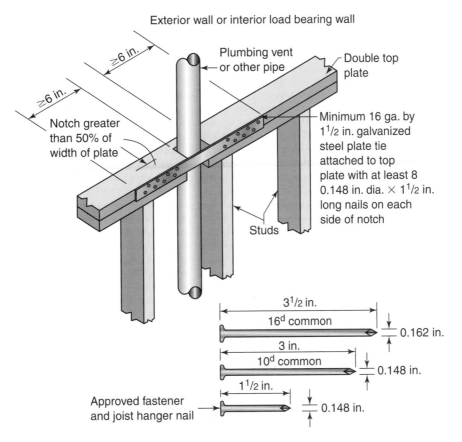

Metal tie required at notched top plate

FIGURE 6-12 Drilling and notching of top plate in exterior wall or bearing interior wall

FIREBLOCKING

To stop the spread of fire in concealed spaces of wood frame construction, fireblocking is required to form an effective barrier between stories and between the top story and the attic. Concealed spaces of stud walls and partitions require fireblocking vertically at the ceiling and floor levels. In platform framing, the top wall plates typically satisfy the fireblocking requirement. The studs generally provide effective fireblocking in the horizontal direction, but for walls with offset studs or other openings, fireblocking is required horizontally at 10-foot intervals or less. Fireblocking also is required at all interconnections between concealed vertical and horizontal spaces, such as those created by soffits, and at the top and bottom of stair stringers (Figure 6-13). Openings around vents, pipes, ducts, cables, and wires must also be sealed at the ceiling and floor level (Figure 6-14). Fireblocking materials include nominal 2-inch-thick lumber, equivalent layers of structural wood panels, and glass fiber insulation securely retained in place. [Ref. R302.11]

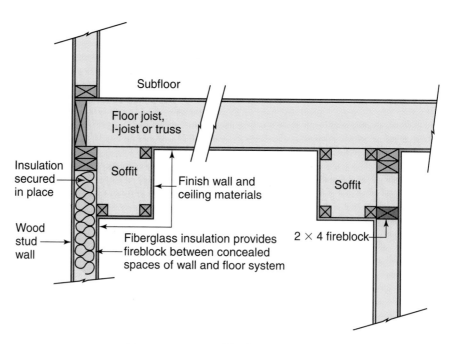

FIGURE 6-13 Fireblocking at soffits

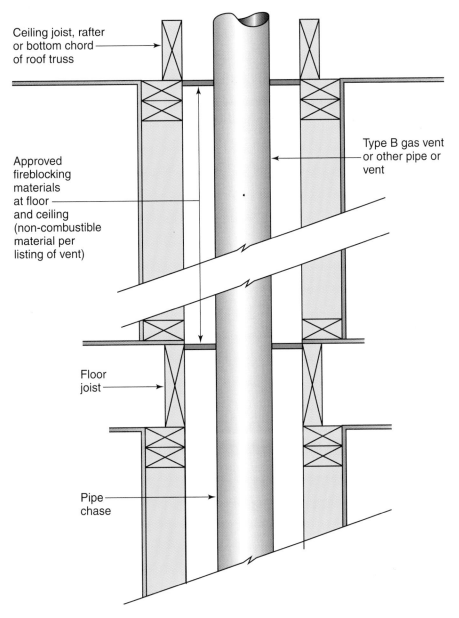

Ceiling joist, rafter or bottom chord of roof truss

Approved fireblocking materials at floor and ceiling (non-combustible material per listing of vent)

Type B gas vent or other pipe or vent

Floor joist

Pipe chase

FIGURE 6-14 Fireblocking at pipe chase

DRAFTSTOPPING

When a ceiling is applied to the bottom side of open web floor trusses, or large areas of communicating spaces are otherwise created in the floor assembly, the IRC requires draftstopping to divide the horizontal spaces into areas of 1000 square feet or less. One-half-inch gypsum board and ⅜-inch wood structural panels are approved draftstopping materials. [Ref. R302.12]

FLOORS

The conventional wood framing floor system includes the support beams or girders, floor joists, floor sheathing, and connections necessary for the load path to the foundation. The IRC provides the lumber framing and wood structural panel spans in prescriptive tables. Floor framing details are shown in Figures 6-15, 6-16, and 6-17.

Beams and girders

The prescriptive spans and support requirements for headers, beams, and girders are based on common species of #2 grade lumber and incorporate the required strength and deflection criteria under code-prescribed uniform loads. **[Ref. Tables R502.5(1) and R502.5(2)]**

> **EXAMPLE**
> Determine the minimum size and bearing support requirements for an interior beam of #2 hem-fir lumber supporting two floors. The width of the building is 28 feet, and the beam span is 6 feet. Refer to Table 6-1 and Figure 6-15.

Joists

In determining the minimum size of a floor joist for a given span, the designer or builder must consider a number of criteria, including live load, dead load, spacing of joists, and the species and grade of lumber. The minimum live load for residential areas is 40 psf, except for sleeping areas and attics served by fixed stairs, where a live load of 30 psf is permitted. The assumed value for uniform dead load in conventional wood frame construction is typically 10 psf. The use of heavier flooring materials such as lightweight concrete or natural stone floor covering may necessitate a joist size based on 20 psf dead load. **[Ref. Tables R502.3.1(1) and R502.3.1(2)]**

> **EXAMPLE**
> Determine the minimum size and maximum spacing of a floor joist of #2 Douglas fir-larch with a span of 14 feet. The floor joists are for a family room, dining room, and kitchen area, and the dead load is 10 psf. Refer to Table 6-2 and Figure 6-16.

Deck attachment

The IRC provides prescriptive methods for attaching a deck to the dwelling to safely resist the applicable loads. The connection details apply to a minimum 2 × 8 preservative treated deck ledger attached to a 2-inch nominal solid-sawn lumber band joist or a minimum 1-inch by 9½-inch Douglas Fir laminated veneer lumber (LVL) rim board. Attachment to other engineered wood products, such as structural composite lumber or wood structural panel band joists requires a design in accordance with accepted engineering practice. Fasteners must be minimum ½-inch diameter hot-dipped galvanized or stainless steel lag screws or bolts

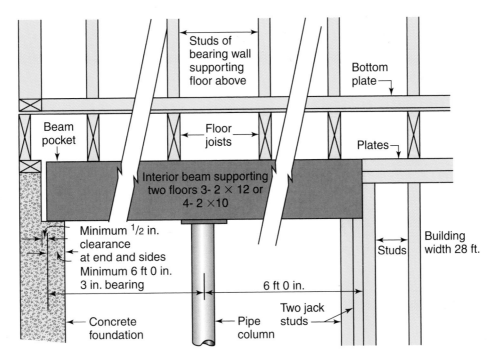

FIGURE 6-15 Interior beam span and bearing support based on Table 6-1

TABLE 6-1 Girder spans and header spans for interior bearing walls (maximum spans for # 2 Douglas fir-larch, hem-fir, southern pine, and spruce-pine-fir, and required number of jack studs)

Headers and Girders Supporting	SIZE	Building Width (feet)					
		20		28		36	
		Span	Jack Studs	Span	Jack Studs	Span	Jack Studs
Two floors	2- 2×4	2–2	1	1–10	1	1–7	1
	2- 2×6	3–2	2	2–9	2	2–5	2
	2- 2×8	4–1	2	3–6	2	3–2	2
	2- 2×10	4–11	2	4–3	2	3–10	3
	2- 2×12	5–9	2	5–0	3	4–5	3
	3- 2×8	5–1	2	4–5	2	3–11	2
	3- 2×10	6–2	2	5–4	2	4–10	2
	3- 2×12	7–2	2	6–3	2	5–7	3
	4- 2×8	6–1	1	5–3	2	4–8	2
	4- 2×10	7–2	2	6–2	2	5–6	2
	4- 2×12	8–4	2	7–2	2	6–5	2

[Ref. Excerpt of Table R502.5(2)]

TABLE 6-2 Floor joist spans for common lumber species, #2 grade (residential living areas, live load = 40 psf, L/360)

Joist Spacing (Inches)	Species and Grade		Dead Load = 10 psf			
			2×6	2×8	2×10	2×12
			Maximum Floor Joist Spans			
			(ft - in.)	(ft - in.)	(ft - in.)	(ft - in.)
12	Douglas fir-larch	#2	10 - 9	14 - 2	17 - 9	20 - 7
	Hem-fir	#2	10 - 0	13 - 2	16 - 10	20 - 4
	Southern pine	#2	10 - 9	14 - 2	18 - 0	21 - 9
	Spruce-pine-fir	#2	10 - 3	13 - 6	17 - 3	20 - 7
16	Douglas fir-larch	#2	9 - 9	12 - 7	15 - 5	17 - 10
	Hem-fir	#2	9 - 1	12 - 0	15 - 2	17 - 7
	Southern pine	#2	9 - 9	12 - 10	16 - 1	18 - 10
	Spruce-pine-fir	#2	9 - 4	12 - 3	15 - 5	17 - 10
19.2	Douglas fir-larch	#2	9 - 1	11 - 6	14 - 1	16 - 3
	Hem-fir	#2	8 - 7	11 - 3	13 - 10	16 - 1
	Southern pine	#2	9 - 2	12 - 1	14 - 8	17 - 2
	Spruce-pine-fir	#2	8 - 9	11 - 6	14 - 1	16 - 3
24	Douglas fir-larch	#2	8 - 1	10 - 3	12 - 7	14 - 7
	Hem-fir	#2	7 - 11	10 - 2	12 - 5	14 - 4
	Southern pine	#2	8 - 6	11 - 0	13 - 1	15 - 5
	Spruce-pine-fir	#2	8 - 1	10 - 3	12 - 7	14 - 7

[Ref. Excerpt of Table R502.3.1(2)]

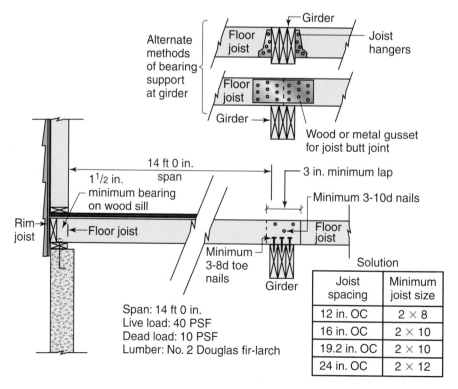

FIGURE 6-16 Floor joist span and bearing details based on Table 6-2

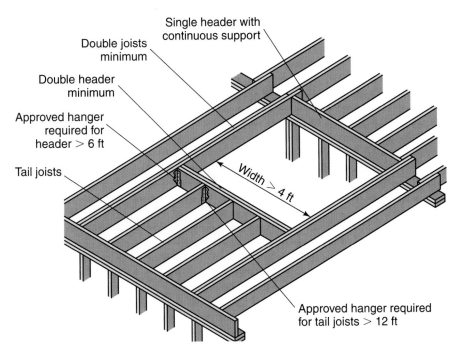

Single header with
continuous support

Double joists
minimum

Double header
minimum

Approved hanger
required for
header > 6 ft

Tail joists

Width > 4 ft

Approved hanger required
for tail joists > 12 ft

FIGURE 6-17 Framing at floor openings

TABLE 6-3 Fastener spacing for a southern pine or hem-fir deck ledger and a 2-inch nominal solid-sawn spruce-pine-fir band joist (deck live load = 40 psf, deck dead load = 10 psf)

Joist Span	6'–0" and Less	6'–1" to 8'–0"	8'–1" to 10'–0"	10'–1" to 12'–0"	12'–1" to 14'–0"	14'–1" to 16'–0"	16'–1" to 18'–0"
Connection Details	**On-Center Spacing of Fasteners**						
1/2" diameter lag screw with 15/32" maximum sheathing	30	23	18	15	13	11	10
1/2" diameter bolt with 15/32" maximum sheathing	36	36	34	29	24	21	19
1/2" diameter bolt with 15/32" maximum sheathing and 1/2" stacked washers	36	36	29	24	21	18	16

[Ref. Table R502.2.2.1]

installed with washers of the same material. The maximum spacing is based on the deck joist span. The code requires a staggered fastener pattern with the bolts or lag screws located 2 inches from the top or bottom of the deck ledger and from 2 to 5 inches from the end of the ledger (Table 6-3 and Figure 6-18). [Ref. R502.2.2.1]

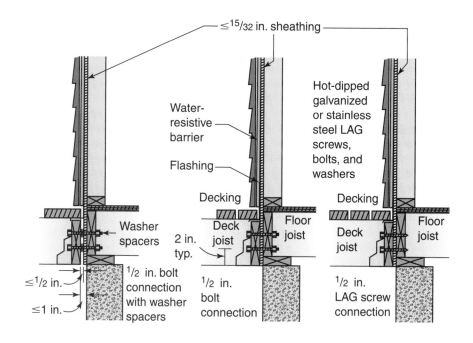

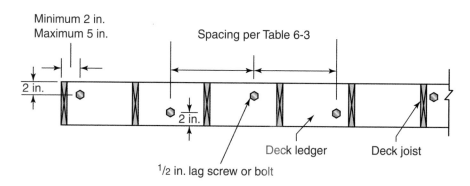

FIGURE 6-18 Deck ledger connection to band joist

WALLS

Walls must be designed and constructed to safely support all code-prescribed loads and transfer those loads to the supporting structure and foundation. In addition to setting limits on the size, length, and spacing of studs, the code includes a number of methods for wall bracing, which is critical to the structural integrity of the building. Wall framing details are shown in Figure 6-19. **[Ref. R602]**

Studs and plates

The prescriptive provisions of the IRC generally limit stud height in bearing walls to 10 feet. The height is the distance between points of lateral support perpendicular to the plane of the wall, which are typically where

EXAMPLE
Determine the minimum size, height, and spacing of standard studs in an exterior bearing wall, as shown in Figure 6-20. Refer to Table 6-4.

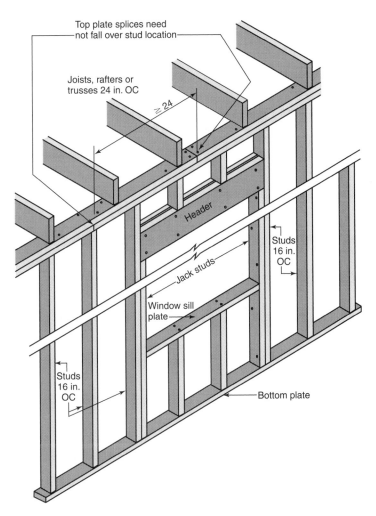

Top plate splices need not fall over stud location

Joists, rafters or trusses 24 in. OC

≥ 24

Header

Studs 16 in. OC

Jack studs

Window sill plate

Studs 16 in. OC

Bottom plate

FIGURE 6-19 Wall framing details

TABLE 6-4 Size, height, and spacing of wood studs

		Bearing Walls				Nonbearing Walls	
Stud Size (inches)	Laterally Unsupported Stud Height (feet)	Maximum Spacing When Supporting a Roof-ceiling Assembly or a Habitable Attic Assembly Only (inches)	Maximum Spacing When Supporting One Floor, Plus a Roof-ceiling Assembly or a Habitable Attic Assembly (inches)	Maximum Spacing When Supporting Two Floors, Plus a Roof-ceiling Assembly or a Habitable Attic Assembly (inches)	Maximum Spacing When Supporting One Floor Only (inches)	Laterally Unsupported Stud Height (feet)	Maximum Spacing (inches)
2 × 3	—	—	—	—	—	10	16
2 × 4	10	24	16	—	24	14	24
3 × 4	10	24	24	16	24	14	24
2 × 5	10	24	24	—	24	16	24
2 × 6	10	24	24	16	24	20	24

[Ref. Excerpt of Table R602.3(5)]

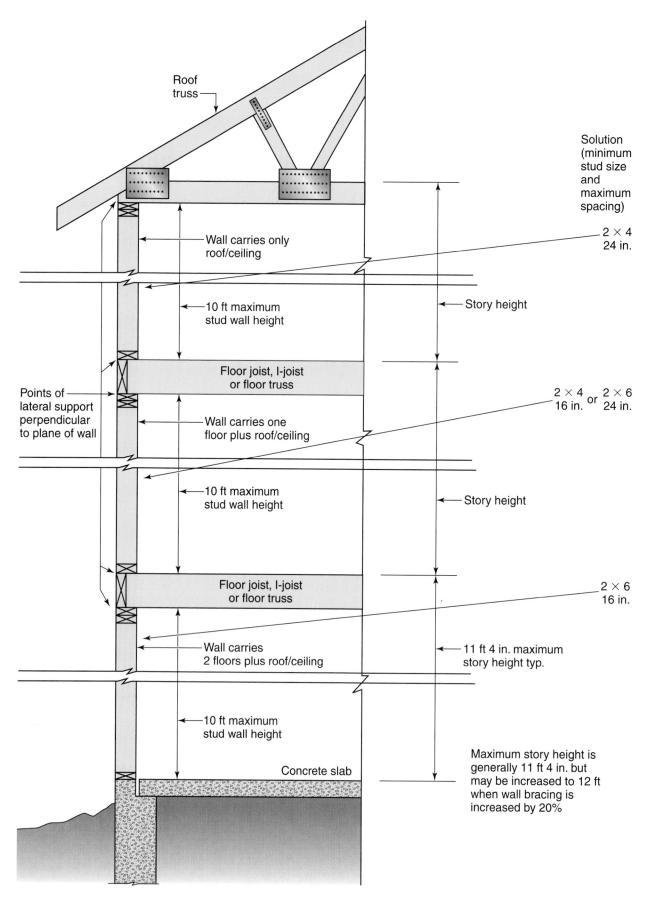

Roof truss

Solution (minimum stud size and maximum spacing)

Wall carries only roof/ceiling

2 × 4
24 in.

10 ft maximum stud wall height

Story height

Floor joist, I-joist or floor truss

Points of lateral support perpendicular to plane of wall

2 × 4 or 2 × 6
16 in. 24 in.

Wall carries one floor plus roof/ceiling

10 ft maximum stud wall height

Story height

Floor joist, I-joist or floor truss

2 × 6
16 in.

Wall carries 2 floors plus roof/ceiling

11 ft 4 in. maximum story height typ.

10 ft maximum stud wall height

Concrete slab

Maximum story height is generally 11 ft 4 in. but may be increased to 12 ft when wall bracing is increased by 20%

FIGURE 6-20 Stud size, height, and spacing based on Table 6-4

the top and bottom plates are connected to the floor or ceiling framing. The size and spacing of studs is related to the number of floors being supported with or without the additional load of the roof-ceiling assembly (Figure 6-20). [Ref. Table R602.3.5]

Headers

Headers are required above door and window openings to carry the loads of construction above and transfer the loads to the wall framing at the sides of the opening. The prescriptive tables for floor girders and beams also provide the span and bearing support requirements for headers. [Ref. Tables R502.5(1) and R502.5(2)]

EXAMPLE

Determine the minimum size and bearing support requirements for a #2 Douglas fir-larch header in an exterior bearing wall as shown in Figure 6-21. The width of the building is 28 feet, the header span is 7 feet, and the snow load is 30 psf. Refer to Table 6-5 and Figure 6-21.

TABLE 6-5 Girder spans and header spans for exterior bearing walls (maximum spans for # 2 grade Douglas fir-larch, hem-fir, southern pine, and spruce-pine-fir and required number of jack studs)

Girders and Headers Supporting	SIZE	Ground Snow Load (psf)			
		30			
		Building Width (feet)			
		28		36	
		Span	Jack Studs	Span	Jack Studs
Roof and ceiling	2- 2×8	5–11	2	5–4	2
	2- 2×10	7–3	2	6–6	2
	2- 2×12	8–5	2	7–6	2
	3- 2×8	7–5	1	6–8	1
	3- 2×10	9–1	2	8–2	2
	3- 2×12	10–7	2	9–5	2
Roof, ceiling, and one center-bearing floor	2- 2×8	5–0	2	4–6	2
	2- 2×10	6–2	2	5–6	2
	2- 2×12	7–1	2	6–5	2
	3- 2×8	6–3	2	5–8	2
	3- 2×10	7–8	2	6–11	2
	3- 2×12	8–11	2	8–0	2
Roof, ceiling, and two center-bearing floors	2- 2×8	4–2	2	3–9	2
	2- 2×10	5–1	2	4–7	3
	2- 2×12	5–10	3	5–3	3
	3- 2×8	5–2	2	4–8	2
	3- 2×10	6–4	2	5–8	2
	3- 2×12	7–4	2	6–7	2

[Ref. Excerpt of Table R502.5(1)]

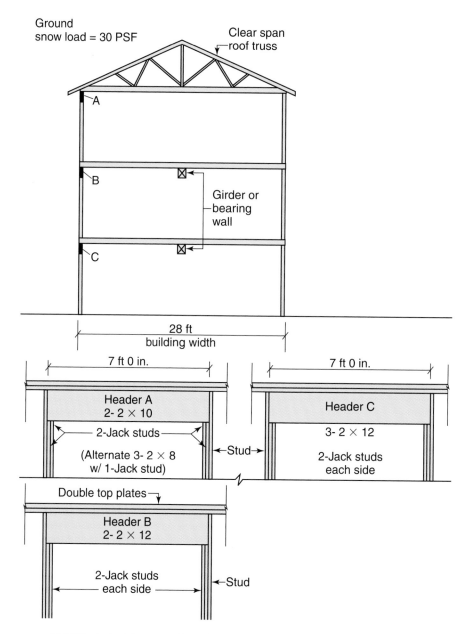

FIGURE 6-21 Exterior header span and bearing support based on Table 6-5

Wall bracing

Wall bracing is necessary to provide resistance to racking from lateral loads, primarily wind and seismic forces. The IRC includes eight distinct and prescriptive methods of panel and diagonal wall bracing, referred to as intermittent bracing methods, though discussion in this chapter focuses on the most common bracing material—wood structural panels (Method WSP). A braced wall panel is the segment of bracing that is the full height of the wall, and its horizontal dimension is referred to as the length of the panel. The minimum length of a braced wall panel is typically 4 feet, but the code offers a number of alternatives to reduce this length by increasing the strength of the panel through specific material, connection, and anchorage details (Figures 6-22, 6-23, and 6-24).

Bracing requirements based on Seismic Design Category				
Seismic Design Category (SDC)				
C	One and two-family dwellings in SDC C are exempt from the seismic requirements—wind controls			
Bracing requirements based on wind speed				
Basic wind speed (mph)	Story location	Braced wall line spacing (feet)	Minimum total length of braced wall panels required along each braced wall line **Method WSP**	Does the 12 feet total length of bracing as shown in the drawing satisfy the bracing length requirements?
≤ 90	One story or second story of 2 stories	50	9	Yes
	First story of 2 stories	30	10.5	Yes
		40	14	No
	First story of 3 stories	20	11	Yes
		30	15.5	No

FIGURE 6-22 Location and minimum length of braced wall panels in a braced wall line

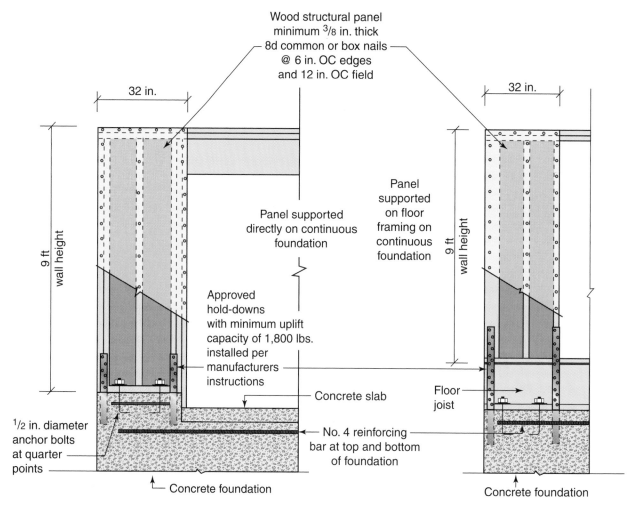

Alternate to replace 48 in. braced wall panel

Wood structural panel minimum 3/8 in. thick 8d common or box nails @ 6 in. OC edges and 12 in. OC field

32 in.

32 in.

9 ft wall height

9 ft wall height

Panel supported directly on continuous foundation

Panel supported on floor framing on continuous foundation

Approved hold-downs with minimum uplift capacity of 1,800 lbs. installed per manufacturers instructions

Concrete slab

Floor joist

1/2 in. diameter anchor bolts at quarter points

No. 4 reinforcing bar at top and bottom of foundation

Concrete foundation

Concrete foundation

FIGURE 6-23 Alternate braced wall panels (Method ABW) in a one-story building with a 9-foot wall height

The series of braced wall panels, typically in an exterior wall, is considered a braced wall line. A braced wall line must contain the prescribed total length of bracing in feet and meet the maximum spacing requirements of braced wall panels as well as braced wall lines. For this reason, interior braced wall lines also may be required. The amount and location of bracing is determined by numerous factors, including the number of stories of the building, the seismic design category, the design wind speed, wind exposure category, and the method of bracing. Another path for compliance with the wall bracing provisions is to apply structural panels to all areas of one side of a braced wall line, including above and below windows. An alternative to intermittent bracing, this continuous sheathing method increases the rigidity of the lateral resistance system and allows reduced lengths for full height braced wall panels (Figure 6-25). **[Ref. R602.10]**

CEILING AND ROOF

The scope of wood roof framing details in the code is limited to roofs with a minimum slope of 3 units horizontal to 12 units vertical (3:12).

A comparison of wall bracing Methods PFG and PFH	
Method PFG—Portal frame at garage door openings in Seismic Design Categories A, B, and C	**Method PFH**—Portal frame with hold-downs
No hold-downs	Hold-downs
Only 1 bottom plate	3 bottom plates
Two ½-inch diameter anchor bolts	One 5/8-inch diameter anchor bolt
2.5-in. x 2.5-in. x 3/16-in. plate washers	3-in. x 3-in. x 0.229-in. plate washer in SDC D_0, D_1, and D_2 and townhouses in SDC C
7/16-inch wood structural panel sheathing	3/8-inch wood structural panel sheathing
Min. 24-inch length of braced wall panel	16-inch BWP length for 1 story 24-inch length for 1st story of 2 stories
Supported directly on continuous foundation	Supported by continuous foundation or by floor framing on continuous foundation

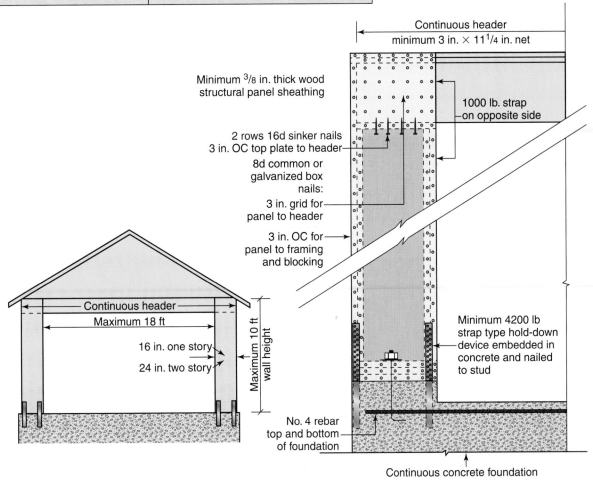

FIGURE 6-24 Portal frame with hold-downs (Method PFH)

Minimum length of braced wall panels (inches)			Adjacent clear opening height (inches)
Wall height			
8 ft	9 ft	10 ft	
35	36	33	84 in.
24	27	30	64 in.

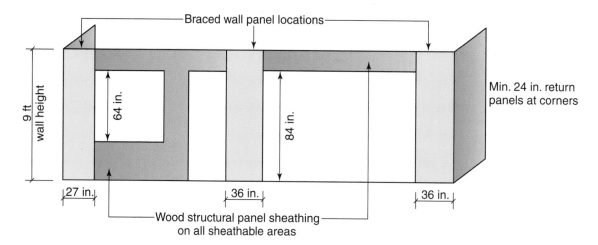

FIGURE 6-25 Continuous wood structural panel sheathing

Ceiling joist

In addition to supporting ceiling materials, ceiling joists also serve as rafter ties to resist the outward thrust of the rafters at the top of the wall. It follows that the ceiling joist requires adequate connection to the rafter, which is in turn fastened to the top of the wall. Maximum ceiling joist spans are provided for attics without storage and attics with limited storage. Attics with fixed stair access require joists sized as floor joists. **[Ref. R802.3 and R802.4]**

Rafters

The prescriptive tables giving maximum spans for rafters are based on the snow load of the geographic area (roof live load of 20 psf is used in areas with snow loads less than 30 psf) and whether the ceiling material is attached to the bottom of the rafter rather than a ceiling joist. Rafters line up opposite each other at the ridge and are typically framed to a ridge board, but gusset plate ties also are permitted. **[Ref. R802.5]**

EXAMPLE

Determine the minimum size and maximum spacing of a #2 spruce-pine-fir rafter with a span of 15 feet, as shown in Figure 6-26. The ground snow load is 30 psf and the dead load is 10 psf. Refer to Figure 6-26 and Table 6-6.

Where ceiling joists are not connected to the rafters at the top plate or are installed perpendicular to the rafters, minimum 2 × 4 rafter ties are required to resist the outward thrust forces of the rafters on the wall. In the absence of joists or rafter ties, the ridge must be supported by a bearing wall or girder, or be designed as a beam (Figures 6-27 and 6-28). **[Ref. R802.3.1]**

Attic ventilation and access

The code requires cross ventilation for each attic or enclosed roof space to prevent moisture from accumulating in the space and causing damage to the structure. In poorly ventilated attics, warm moist air escaping from the conditioned space condenses on the framing and sheathing of the cooler attic space. The total net free ventilating area must be at least

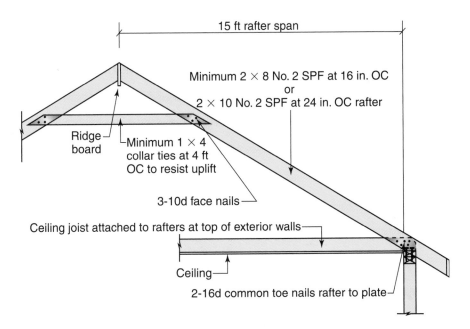

FIGURE 6-26 Rafter span with no ceiling attached based on Table 6-6

TABLE 6-6 Rafter spans for common lumber species, # 2 grade (ground snow load = 30 psf, ceiling not attached to rafters, L/180, dead load = 10 psf)

Rafter Spacing (inches)	Species and Grade		Dead Load = 10 psf				
			2 × 4	2 × 6	2 × 8	2 × 10	2 × 12
			Maximum Rafter Spans				
			ft. - in.	ft. - in.	ft. - in.	ft. - in.	ft. - in.
16	Douglas fir-larch	#2	8-2	11-11	15-1	18-5	21-5
	Hem-fir	#2	8-0	11-9	14-11	18-2	21-1
	Southern pine	#2	8-7	12-6	16-2	19-3	22-7
	Spruce-pine-fir	#2	8-2	11-11	15-1	18-5	21-5
19.2	Douglas fir-larch	#2	7-5	10-11	13-9	16-10	19-6
	Hem-fir	#2	7-4	10-9	13-7	16-7	19-3
	Southern pine	#2	7-11	11-5	14-9	17-7	20-7
	Spruce-pine-fir	#2	7-5	10-11	13-9	16-10	19-6
24	Douglas fir-larch	#2	6-8	9-9	12-4	15-1	17-6
	Hem-fir	#2	6-7	9-7	12-2	14-10	17-3
	Southern pine	#2	7-1	10-2	13-2	15-9	18-5
	Spruce-pine-fir	#2	6-8	9-9	12-4	15-1	17-6

[Ref. Excerpt of Table R802.5.1(3)]

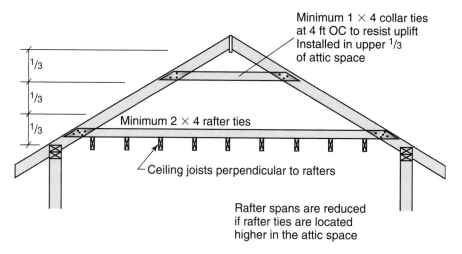

FIGURE 6-27 Rafter ties

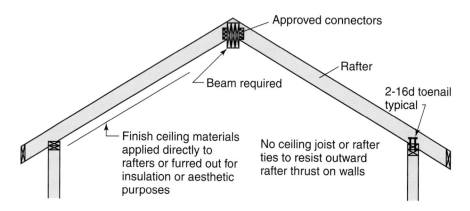

FIGURE 6-28 Ridge beam

$1/150$ of the area of the space. A reduction to $1/300$ the area of the space is permitted when 50 to 80 percent of the required ventilating area is in the upper portion of the space, with the balance of the ventilating area provided by eave or cornice vents. Unvented attics are permitted under certain conditions. **[Ref. R806]**

Access to attics is required when the attic area exceeds 30 square feet and has a height of 30 inches or greater. The access requires a rough opening of at least 22 by 30 inches with headroom above the opening at least 30 inches high. The access opening must be located in a hallway or other readily accessible location. **[Ref. R807]**

PART IV

Finishes and Weather Protection

Chapter 7: Interior and Exterior Finishes and Weather Protection

Interior and Exterior Finishes and Weather Protection

As part of its purpose statement to protect the health and general welfare of the public, the *International Residential Code* (IRC) sets minimum requirements for durable interior and exterior finishes. Exterior wall and roof coverings protect the structure against the damaging effects of water intrusion and provide an envelope for a healthy and livable interior environment.

INTERIOR FINISHES

The IRC includes minimum installation requirements for gypsum board (drywall), plaster, ceramic tile, and wood paneling for walls and ceilings. Inspection is not required for other than attachment of lath or gypsum board that is part of a fire resistant rated assembly. Fire resistance requirements for walls and ceilings are covered in Chapter 9. Interior finishes require protection from moisture and are typically installed at a stage of construction when the building is substantially weather tight. The IRC does not regulate the installation of floor coverings or the application of paint and wallpaper. [Ref. R702]

Gypsum board

Gypsum board is the generic term for sheet panel products with a non-combustible gypsum core and paper facing that is commonly used for the wall and ceiling finish of dwelling construction. When taped and finished, it is often referred to as drywall. Type X gypsum board contains core additives for greater fire resistance than regular gypsum board. Gypsum products may also be reinforced for greater strength or manufactured for greater water resistance or durability.

The minimum fastening and thickness requirements for gypsum board relate to the spacing of the framing members and the location and intended application of the gypsum board. Generally, gypsum board may be installed with the long dimension perpendicular or parallel to framing members. The code requires installation perpendicular to ceiling framing members for ⅜-inch material and places additional limitations on ceiling applications receiving water-based textures or serving as a fire resistant rated separation. Tables 7-1 and 7-2 summarize gypsum board application requirements. [Ref. R702.3, Table R702.3.5]

Backing for ceramic tile and other nonabsorbent finishes

To prevent deterioration of the substrate and damage to the wall structure, only fiber-cement, fiber-mat-reinforced cement, glass mat gypsum backers, or fiber-reinforced gypsum backers are permitted as backing material for wall tile installed in tub and shower areas. This requirement also applies to backers for nonabsorbent plastic panels in showers. In addition, installation of water resistant gypsum backing board is not allowed where directly exposed to water or in areas subject to continuous high humidity. Water resistant gypsum backing board is permitted on ceilings where framing spacing does not exceed 12 inches on center (O.C.) for ½-inch material or 16 inches O.C. for ⅝-inch material. [Ref. R702.3.8 and R702.4]

EXTERIOR WALL COVERINGS

Water resistant barriers, flashing, windows, doors, and siding or veneers form the protective exterior envelope of a dwelling. [Ref. R703]

TABLE 7-1 Minimum thickness and application of gypsum board

Thickness of Gypsum Board (in.)	Orientation of Gypsum Board to Framing	Maximum Spacing of Framing Members (in. O.C.)	Maximum Spacing of Fasteners (in.)	
			Nails	Screws
Wall Application				
3/8	Either direction	16	8	16
1/2	Either direction	24	8	12
		16	8	16
5/8	Either direction	24	8	12
		16	8	16
Ceiling Application, No Water-Based Texture Material				
3/8*	Perpendicular	16	7	12
1/2	Either direction	16	7	12
	Perpendicular	24	7	12
5/8	Either direction	16	7	12
	Perpendicular	24	7	12
Ceiling Application with Water-Based Texture Material				
1/2	Perpendicular	16	7	12
1/2 sag-resistant	Perpendicular	24	7	12
5/8	Perpendicular	16	7	12
	Perpendicular	24	7	12
Garage Ceiling Application with Habitable Space Above				
5/8 Type X	Perpendicular	16	6	6
	Perpendicular	24	6	6

*Not permitted to support insulation. O.C. = on center.

TABLE 7-2 Fasteners for the application of gypsum board

Thickness of Gypsum Board (in.)	Screws		Nails			
	Attached to Steel Framing	Attached to Wood Framing	Attached to Wood Framing			
	Type S	Type W or Type S	13 Gauge	Ring-Shank	Cooler	Gypsum Board Nail
	Minimum Length (in.)					
3/8	3/4	1	1 1/4	1 1/4	1 3/8	
1/2	7/8	1 1/8	1 3/8	1 1/4	1 5/8	1 5/8
5/8	1	1 1/4	1 5/8	1 3/8	1 7/8	1 7/8

Water and moisture management

In wood or steel light-frame construction, for other than detached accessory buildings, the code requires a water resistant barrier over the sheathing of all exterior walls. Siding and veneers are typically not impervious to wind-driven rain, and the water-resistive barrier in combination with flashings completes the weather protective system to keep moisture out of the wall assembly. The IRC prescribes one layer of No. 15 asphalt felt applied horizontally with 2-inch laps for the water-resistive barrier, but approved house wrap and other materials tested to perform equivalently to the felt satisfy the requirement. House wraps must be installed in accordance with the manufacturer's instructions to shed water away from the sheathing to the outside of the wall coverings. **[Ref. R703.1 and 703.2]**

Flashing

To prevent water entering behind exterior wall coverings and penetrating the wall assembly, the code requires corrosion-resistant flashing at specific locations, including exterior window and door openings, penetrations, projections, wall and roof intersections, and intersections of dissimilar materials. Flashing is particularly important for the ledger attachment joining a porch or deck to the house structure and protects not only the concealed framing but the structural integrity of the deck or porch (Figure 7-1). **[Ref. R703.8]**

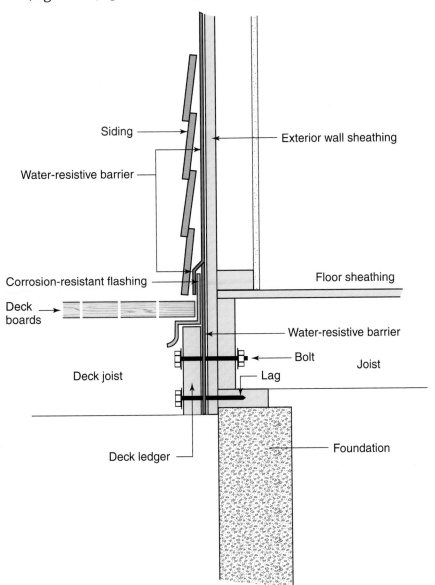

FIGURE 7-1 Wall flashing at deck

Masonry and stone veneer

Because of its vulnerability to damage due to earthquake forces, the prescriptive provisions of the IRC place greater limitations on the height, thickness, and weight of masonry and stone veneers in the higher Seismic Design Categories (SDCs). For buildings sited in SDC "A," "B," or "C," the code generally permits veneers up to three stories and 30 feet above noncombustible foundations, with an additional 8 feet for gable end walls. Maximum thickness is 5 inches and the maximum weight is 50 psf. The code reduces the maximum height, thickness, and weight of veneer for buildings located in SDC "D_0", "D_1", or "D_2". See Figure 7-2 for typical masonry veneer details. [**Ref. R703.7**]

Support

Masonry veneer typically is supported by a continuous concrete or masonry foundation. In SDC "A," "B," or "C," light-frame construction may support exterior veneer weighing not more than 40 psf when designed to limit deflection to $^1/_{600}$ of the span of the supporting members. Steel or noncombustible lintels are required above openings and must have bearing support of at least 4 inches at each end. Steel lintels require a

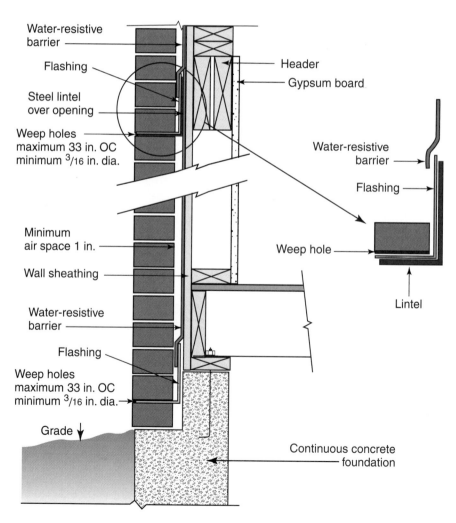

FIGURE 7-2 Brick veneer details

rust-inhibitive shop coat on all surfaces or otherwise be protected against corrosion (Table 7-3, Figure 7-3). [Ref. R703.7.3]

Veneer anchoring

Veneer is anchored to the structure with corrosion-resistant metal ties of No. 9-gauge strand wire or No. 22-gauge × ⅞-inch corrugated sheet metal. The number of anchors is increased for installations in SDCs "D_0," "D_1," and "D_2" and for townhouses in SDC "C" (Figure 7-4). [Ref. R703.7.4]

TABLE 7-3 Allowable spans for steel lintels supporting masonry veneer

Size of Steel Angle (in.)	No Story Above	One Story Above	Two Stories Above
3 × 3 × ¼	6'-0"	4'-6"	3'-0"
4 × 3 × ¼	8'-0"	6'-0"	4'-6"
5 × 3½ × ⁵⁄₁₆	10'-0"	8'-0"	6'-0"
6 × 3½ × ⁵⁄₁₆	14'-0"	9'-6"	7'-0"
2 − 6 × 3½ × ⁵⁄₁₆	20'-0"	12'-0"	9'-6"

Note: Long leg of the angle shall be placed in a vertical position.

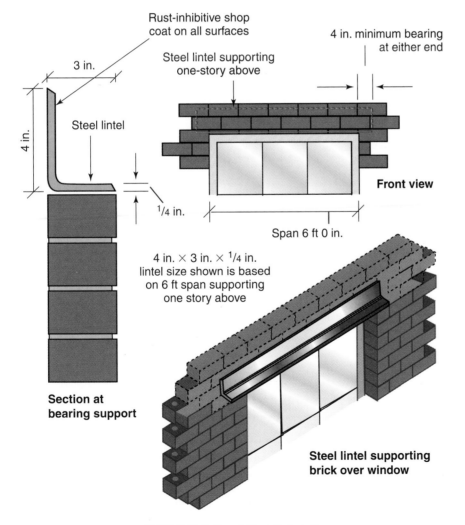

Rust-inhibitive shop coat on all surfaces

Steel lintel supporting one-story above

4 in. minimum bearing at either end

3 in.

Steel lintel

4 in.

¼ in.

Front view

Span 6 ft 0 in.

4 in. × 3 in. × ¼ in. lintel size shown is based on 6 ft span supporting one story above

Section at bearing support

Steel lintel supporting brick over window

FIGURE 7-3 Steel lintel details

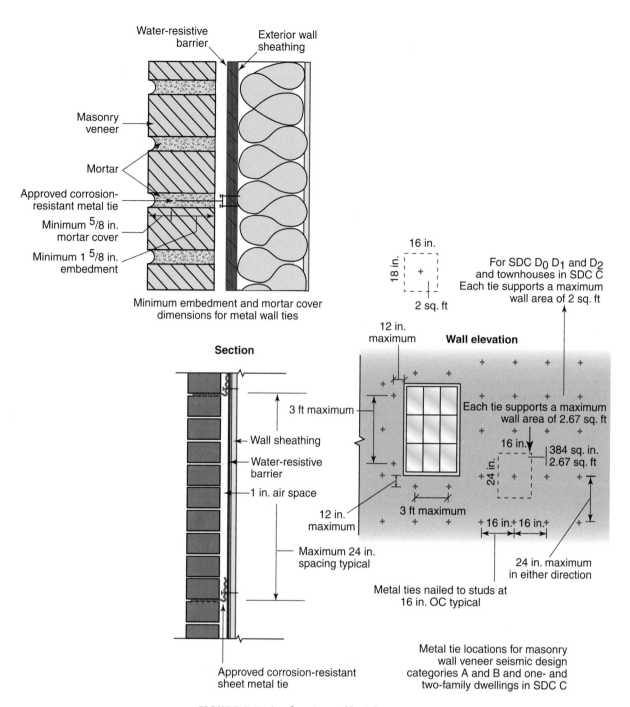

FIGURE 7-4 Anchoring of brick veneer

Siding

Vertical panel siding and horizontal lap siding must be installed with proper joint treatments and flashing to resist penetration of moisture. Vertical joints require batten covers or a combination of flashing and sealants. Horizontal joints of panel siding require a minimum 1 inch lap, shiplap or a "Z" flashing over solid backing. Horizontal lap siding must be installed in accordance with the manufacturer's recommendations. In the absence of recommendations, the code requires a minimum lap of 1 inch to 1¼ inches depending on the siding material. All siding requires secure attachment with approved corrosion-resistant fasteners, which

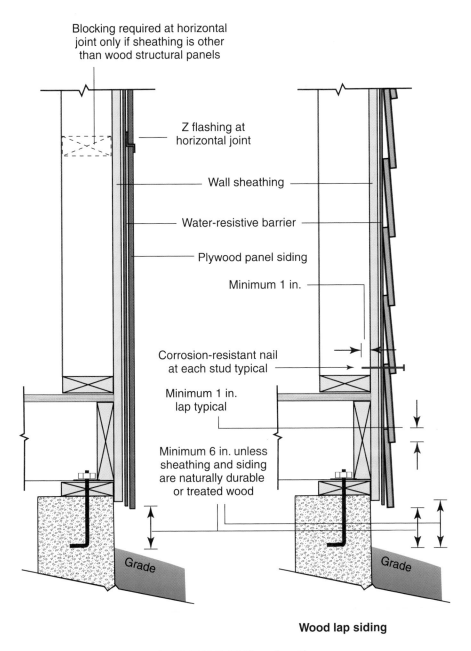

Blocking required at horizontal
joint only if sheathing is other
than wood structural panels

Z flashing at
horizontal joint

Wall sheathing

Water-resistive barrier

Plywood panel siding

Minimum 1 in.

Corrosion-resistant nail
at each stud typical

Minimum 1 in.
lap typical

Minimum 6 in. unless
sheathing and siding
are naturally durable
or treated wood

Grade

Grade

Wood lap siding

FIGURE 7-5 Siding details

generally must penetrate into the wood framing 1 to 1½ inches, depending on the siding material and the manufacturer's recommendations (Figure 7-5). Vinyl siding must be installed in accordance with the manufacturer's installation instructions and meet other applicable requirements based on design wind speed and wind exposure category. [Ref. R703.3, R703.4, R703.10, R703.11, and Table R703.4]

Exterior insulation finish system (EIFS)

EIFS, sometimes referred to as synthetic stucco, due to its textured appearance, consists of a water-resistive barrier, rigid polystyrene insulation, a reinforced base coat, and a trowel-applied textured finish coat. The code requires a means to drain any moisture trapped behind the EIFS to the exterior. To protect the integrity of the weather repellant surface

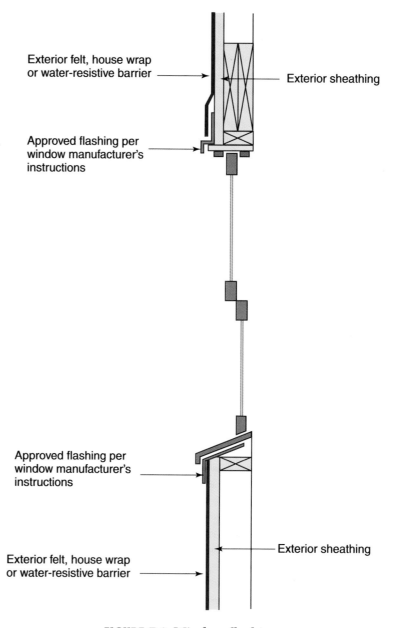

Exterior felt, house wrap or water-resistive barrier

Exterior sheathing

Approved flashing per window manufacturer's instructions

Approved flashing per window manufacturer's instructions

Exterior sheathing

Exterior felt, house wrap or water-resistive barrier

FIGURE 7-6 Window flashing

and prevent moisture penetration, face nailing of trim through the EIFS is not permitted. A minimum of 6 inches clearance is required between the ground and the lowest edge of the EIFS. In addition to the above code requirements, all EIFS must be installed in accordance with the manufacturer's installation instructions. **[Ref. R703.9]**

Windows

In the installation of windows, the code is concerned with the strength of the unit and its attachment to the structure to resist applicable wind loads, as well as the water resistance of the installation in the wall assembly. In this regard, windows must be manufactured to comply with the applicable referenced standards, and installation must follow the manufacturer's written instructions. The IRC also requires the manufacturer to provide installation and flashing detail instructions with each window (Figure 7-6). **[Ref. R612 and R703.8]**

ROOF COVERING

The IRC prescribes the design, materials, construction, and quality of roofing assemblies to provide weather protection for the building. Roof coverings must be installed according to the code and the manufacturer's instructions. This section will focus on the installation of asphalt and wood shingles, as well as the associated underlayment and flashing requirements.

Underlayment and ice barrier

At a minimum, for slopes of 4:12 or greater, one layer of No. 15 asphalt-saturated organic felt or other approved material is required to cover the roof deck before application of shingles. The code requires horizontal laps of at least 2 inches with end laps offset at least 6 feet in successive courses of felt (Figure 7-7). Slopes of at least 2:12 and less than 4:12 require two layers of felt with 19-inch horizontal overlaps (Figure 7-8). [Ref. R905.2.7]

In cold climates, ice dams often form in gutters and along eaves. Freeze-thaw cycles combined with warm air escaping from the conditioned space thaws ice above the eave area. Water unable to drain past the ice dam is often forced back under the shingles and underlayment, causing damage to the structure beneath. In areas with a history of water damage to structures from ice dams at roof eaves, an ice barrier is

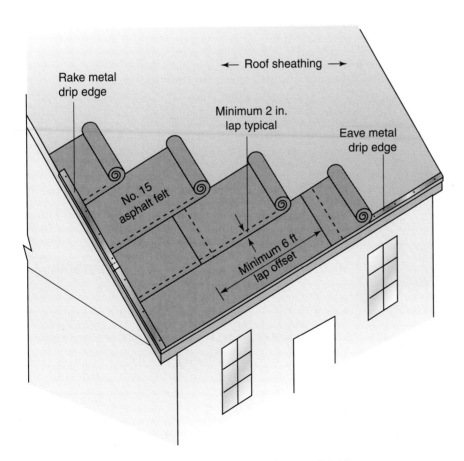

FIGURE 7-7 Roof underlayment for slopes of 4:12 or greater

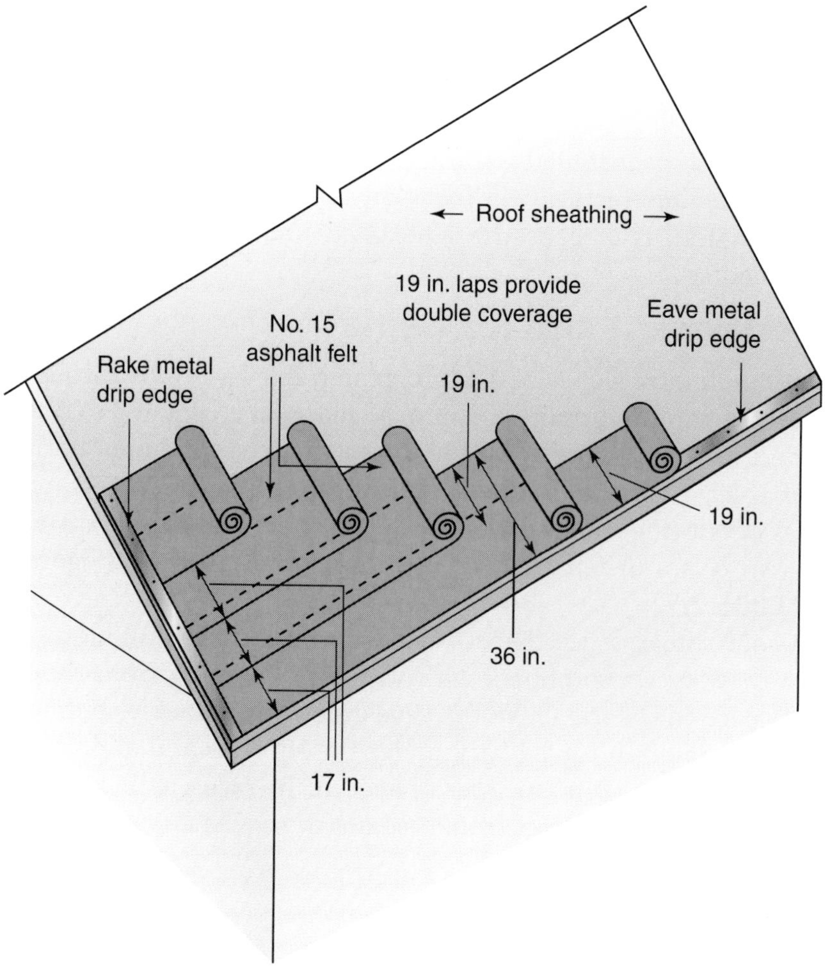

FIGURE 7-8 Roof underlayment for low slope application, slopes 2:12 or greater and less than 4:12

required for added protection. The ice barrier consists of self-adhering polymer-modified bitumen material or two layers of cemented underlayment, which must extend from the eave to at least 24 inches inside the exterior wall line of the building (Figure 7-9). As discussed in Chapter 4, the jurisdiction indicates the ice barrier requirement in the "Climatic and Geographic Design Criteria" table when adopting the IRC. As damage from ice damming is less likely to occur, ice barriers are not required in unheated detached accessory buildings. [Ref. R905.2.7.1 and R905.7.3.1]

Flashing

To effectively seal against entry of water, flashing is required at roof/wall intersections, at points of change in slope or direction, and around roof openings or penetrations. Where the roof joins a sidewall, step-type flashing is required (Figure 7-10). The code requires flashing to be corrosion-resistant metal at least 0.019 inch thick (No. 26 galvanized sheet). Any chimney penetration more than 30 inches wide requires a cricket or saddle to divert water from the roof above to each side of the chimney. Crickets may be the same material as the roof covering or sheet metal (see Chapter 11). [Ref. R905.2.8]

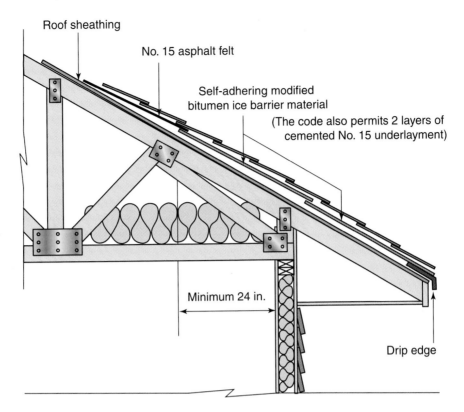

Roof sheathing

No. 15 asphalt felt

Self-adhering modified
bitumen ice barrier material

(The code also permits 2 layers of
cemented No. 15 underlayment)

Minimum 24 in.

Drip edge

FIGURE 7-9 Ice barrier

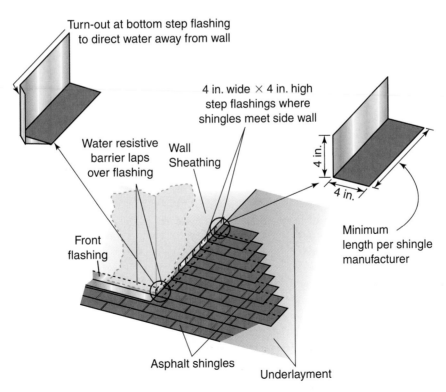

Turn-out at bottom step flashing
to direct water away from wall

4 in. wide × 4 in. high
step flashings where
shingles meet side wall

Water resistive
barrier laps
over flashing

Wall
Sheathing

4 in.

4 in.

Minimum
length per shingle
manufacturer

Front
flashing

Asphalt shingles

Underlayment

Sidewall flashing for asphalt shingles

FIGURE 7-10 Sidewall step flashing

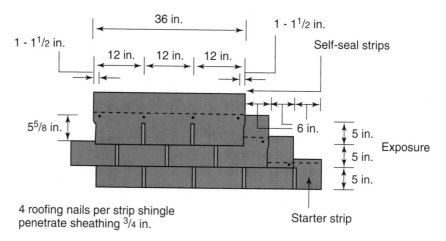

FIGURE 7-11 Asphalt shingles

Asphalt shingles

Asphalt shingles require a roof slope of at least 2:12 and must be installed in accordance with the manufacturer's instructions. Adequate wind resistance is addressed through testing to the appropriate standard for self-sealing or unsealed shingles and labeling to indicate the applicable classification for the design wind speed of the geographic location. Fasteners must be galvanized steel, stainless steel, aluminum, or copper roofing nails of at least 12 gauge (0.105 inch) with a head diameter not less than ⅜ inch. Nails must penetrate at least ¾ inch into the roof sheathing or penetrate through the sheathing (Figure 7-11). **[Ref. R905.2]**

The code recognizes a number of accepted practices for valley construction for asphalt shingles. Closed valleys (covered with shingles) require a valley lining consisting of a self-adhering polymer-modified bitumen sheet, an approved 24-inch-wide metal valley, one ply of approved smooth roll roofing at least 36 inches wide or two plies of mineral-surfaced roll roofing. Open valleys consist of an approved 24-inch-wide metal valley or two plies of mineral-surfaced roll roofing (Figure 7-12). **[Ref. R905.2.8.2]**

Wood shingles and wood shakes

The IRC includes requirements specific to wood shingles and shakes used for roofing. In addition to material and grading requirements, the code provides installation details related to slope, decking, underlayment, laps, exposure, and fastening. The minimum slope is 3:12. Valleys require at least No. 26-gauge corrosion-resistant sheet metal extended 10 inches from the centerline each way for wood shingles and 11 inches from the centerline each way for shakes. End laps must be at least 4 inches (Tables 7-4 and 7-5 and Figures 7-13 and 7-14). **[Ref. R905.7 and R905.8]**

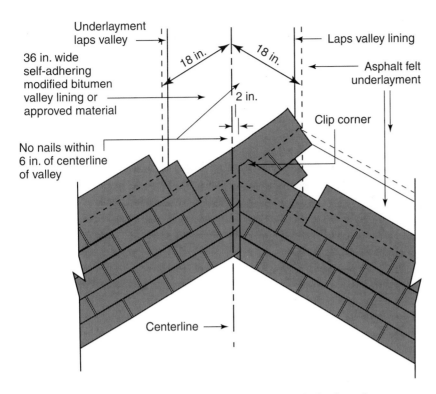

FIGURE 7-12 Cut closed valley for asphalt shingles

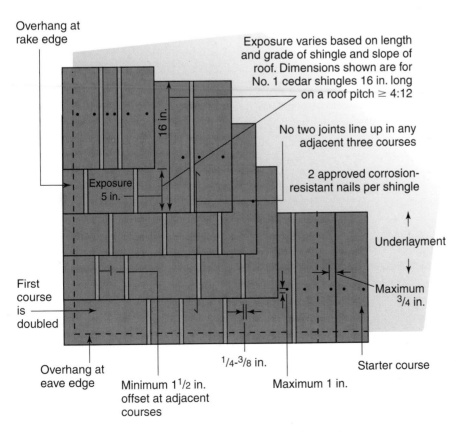

FIGURE 7-13 Wood shingles

You Should Know

The difference between wood shingles and shakes

Wood shingles

- Sawn on both sides
- Tapered
- Approximately ⅜ inch thick at butt

Wood shakes

- Split on one or both sides, or may be sawn on both sides
- Approximately ½ inch or ¾ inch thick at butt
- Usually tapered
- Greater texture and shadow lines than wood shingles ●

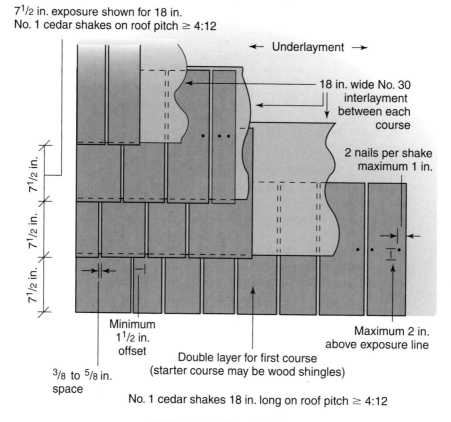

7½ in. exposure shown for 18 in.
No. 1 cedar shakes on roof pitch ≥ 4:12

← Underlayment →

18 in. wide No. 30 interlayment between each course

2 nails per shake maximum 1 in.

7½ in.

7½ in.

7½ in.

7½ in.

Minimum 1½ in. offset

³/₈ to ⁵/₈ in. space

Double layer for first course (starter course may be wood shingles)

Maximum 2 in. above exposure line

No. 1 cedar shakes 18 in. long on roof pitch ≥ 4:12

FIGURE 7-14 Wood shakes

TABLE 7-4 Wood shingle weather exposure and roof slope

Roofing Material	Length (in.)	Grade	Exposure (in.)	
			3:12 Pitch to <4:12	4:12 Pitch or Steeper
Shingles of naturally durable wood	16	No. 1	3¾	5
		No. 2	3½	4
		No. 3	3	3½
	18	No. 1	4¼	5½
		No. 2	4	4½
		No. 3	3½	4
	24	No. 1	5¾	7½
		No. 2	5½	6½
		No. 3	5	5½

TABLE 7-5 Wood shake weather exposure and roof slope

Roofing Material	Length (in.)	Grade	Exposure (in.), 4:12 Pitch or Steeper
Shakes of naturally durable wood	18	No. 1	7½
	24	No. 1	10*
	18	No. 1	7½
Preservative-treated taper-sawn shakes of southern yellow pine	24	No. 1	10
	18	No. 2	5½
	24	No. 2	7½
	18	No. 1	7½
Taper-sawn shakes of naturally durable wood	24	No. 1	10
	18	No. 2	5½
	24	No. 2	7½

*For 24-in. by ⅜-in. handsplit shakes, the maximum exposure is 7½ in.

Reroofing

Where existing roofing is water soaked, deteriorated, or such that it is not adequate as a base for additional roofing, existing roof coverings must be removed before applying new roofing. Wood shake, slate, clay, cement, asbestos-cement tile, or two layers of any roofing material are also not suitable for a roof overlay and must be removed. In addition, asphalt shingles in an area subject to moderate or severe hail exposure are not permitted without first removing existing roof coverings. [Ref. R907]

PART V

Health and Safety

Home Safety

In protecting the health, safety, and welfare of the dwelling occupants, the *International Residential Code* (IRC) sets minimum requirements for a safe means of exiting the building, protection from falls (Figure 8-1) and from the hazards associated with breaking glass, and room dimensions to support a healthy living environment.

FIGURE 8-1 Handrail and guard

ROOM AREAS

Overcrowding in confined spaces creates unhealthy and unsafe living conditions. Though most homes will far exceed the minimum requirements, the IRC recognizes the need for basic living spaces. The code requires habitable rooms other than kitchens to be 70 square feet or larger, with the smallest dimension no less than 7 feet. At least one habitable room of a dwelling must be not less than 120 square feet. **[Ref. R304]**

CEILING HEIGHT

Adequate ceiling height contributes to a healthy living environment and provides the ability to move about and safely exit the building. The general rule establishes a minimum ceiling height of 7 feet for all usable spaces of a dwelling other than closets (Figures 8-2 and 8-3). The code allows for sloped ceilings, provided that half of the required room area accommodates the 7-foot height. Reductions are also permitted in basements without habitable spaces. **[Ref. R305]**

MEANS OF EGRESS

Means of egress describes the path of travel from any location in the dwelling to the exterior. The IRC regulates stairways, ramps, hallways, and doors as the primary components of that path for a safe exit from the building. Hallways must have a clear width of 3 feet; and one exterior exit door is required, also 3 feet in width. Otherwise, the IRC does not regulate the size or type of doors or limit the travel distance from any portion of the dwelling unit to the required exit. As a measure for protecting the path for safe exit from the building, the code requires limited fire protection on the

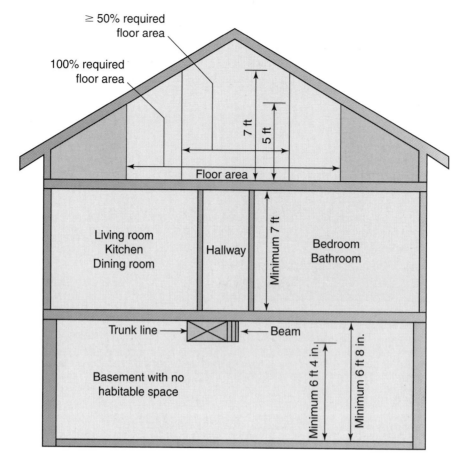

FIGURE 8-2 Ceiling height

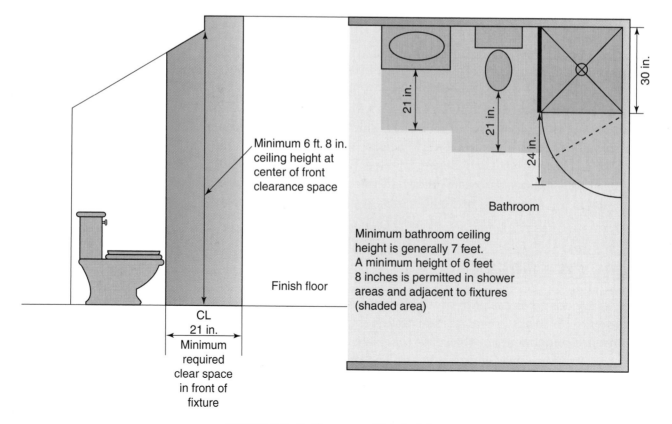

FIGURE 8-3 Bathroom ceiling height

underside of stairs by applying ½-inch gypsum board on the enclosed side and emphasizes that elements on the exterior side of an exit door, such as a deck or porch, must be securely anchored to the structure. Another important element along the path of egress is the ability of the occupant to open the required exit door without a key or special knowledge. This precludes the use of a double-keyed deadbolt. [Ref. R311]

Doors and landings

For each dwelling unit, the code requires one side-hinged exterior exit door providing a net opening of 32 inches by 78 inches, typically achieved with the installation of a door with nominal measurements of 3 feet by 6 feet 8 inches. Occupants in any location of the dwelling must be provided a route to the required exit door without passing through a garage. A landing or floor is generally required on each side of exterior doors with a maximum threshold height above the landing of 1½ inches. An exception allows the exterior landing at the required exit door to be not more than 7¾ inches below the top of the threshold, provided the door swings in. At other than the required exit door, the floor or landing on either side of the door is permitted to be 7¾ inches below the top of the threshold, and the door may swing in either direction (Figures 8-4 and 8-5). The code requires landings to be at least as wide as the door and not less than 36 inches in the direction of travel. A stair without a landing is permitted outside a door other than the required exit door if the door swings in and the stair has only two risers (Figure 8-6). [Ref. R311.3]

Stairs

Fall injuries on stairs are not uncommon, and the code endeavors to improve stair safety through proper pitch, walking surface, clearances, uniformity, and graspable handrails. The minimum 10-inch treads and maximum 7¾-inch risers determine the maximum steepness of the stairway, but just as important in stair safety is the uniformity of those treads and risers for the full flight of the stair. As a person walks a stair, he or she anticipates that the next step will be the same as the previous one. Variations that are not visually apparent may break the user's rhythm or otherwise cause a misstep and fall (Figures 8-7 through 8-9). [Ref. R311.7]

Winders

Winder treads have nonparallel edges, and the code permits a tread depth of 6 inches at the narrow end, provided the full tread depth of 10 inches is achieved within 12 inches of the narrow side (Figure 8-10). A person walking on winder treads and holding the handrail will typically be positioned 12 inches from the narrow side (referred to as the "walk line"), and this configuration allows a turn in the stairway without a landing and without creating an undue hazard. [Ref. R311.7.4.2]

Spiral stairways

The code permits spiral stairways at any location in a dwelling (Figure 8-11). They have lesser dimensions for width, treads, and headroom and greater riser heights than conventional stairs. A spiral stairway may serve as a means of egress from a story. [Ref. R311.7.9.1]

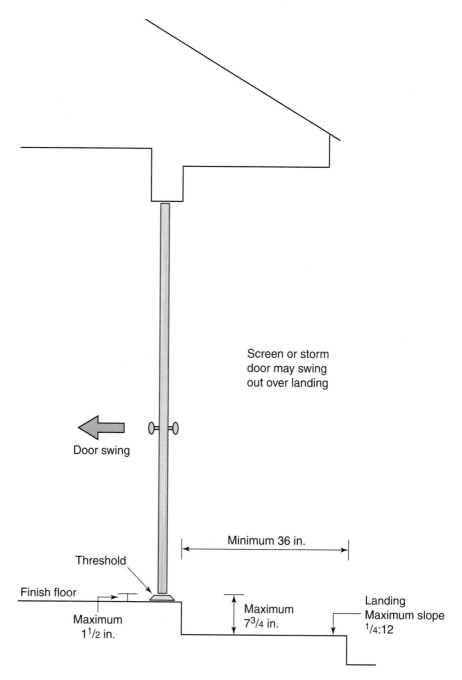

Screen or storm
door may swing
out over landing

Door swing

Minimum 36 in.

Threshold

Finish floor

Maximum
1¹/₂ in.

Maximum
7³/₄ in.

Landing
Maximum slope
¹/₄:12

FIGURE 8-4 Landing at required exterior exit door

Stair landings

Similar to the general rule requiring landings at exterior doors, in most cases a floor or landing is required at the top and bottom of stairs. This is usually not an issue unless a door is installed to enclose the stairway or the stairway opening is framed too close to a wall. The landing requirement prevents the installation of a door in close proximity to the bottom tread. Such an installation would create not only a headroom problem but a falling hazard as well. An exception to the landing requirement allows a door at the top of an interior flight of stairs, provided the door does not swing over the step (Figure 8-12). **[Ref. R311.7.5]**

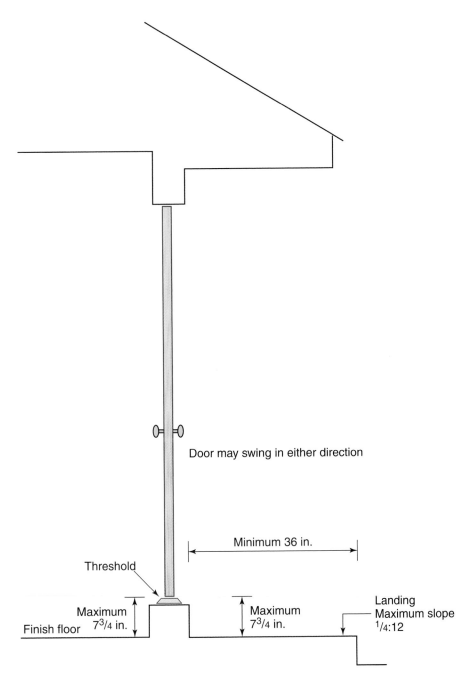

Door may swing in either direction

Minimum 36 in.

Threshold

Maximum 7³/4 in.

Finish floor

Maximum 7³/4 in.

Landing
Maximum slope
¹/4:12

FIGURE 8-5 Landing and floor at exterior door that is not the required exit

Handrails

Handrails are a critical component of stair safety. To be effective, they must be placed 34 to 38 inches above the tread nosing, be continuous, and have a shape that is easily grasped and held (Figures 8-13 and 8-14). The cross section of circular handrails requires a diameter of 1¼ to 2 inches. The code provides dimensions for other than round handrails, but any shape, type, or size of handrail that provides equivalent graspability is acceptable. This is a performance criterion, and the building official holds the responsibility for determining equivalency. Handrails must also be securely anchored to resist a single concentrated load of 200 pounds applied in any direction. [Ref. R311.7.7]

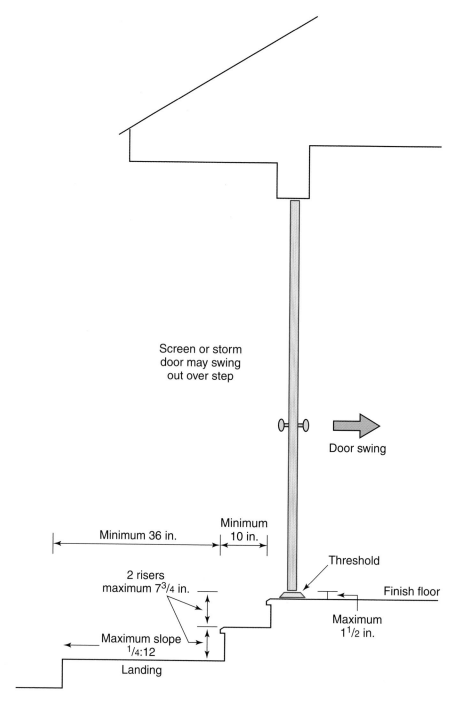

FIGURE 8-6 Steps at exterior door that is not the required exit

PROTECTION FROM FALLS

The IRC intends to protect dwelling occupants from fall injuries at prescribed locations considered hazardous by regulating the design and installation of guards and the height of window sills.

Guards

The IRC generally requires a minimum 36-inch-high guard as protection against falling from a walking surface to a lower surface more than 30 inches below. In determining where a guard is required, the vertical

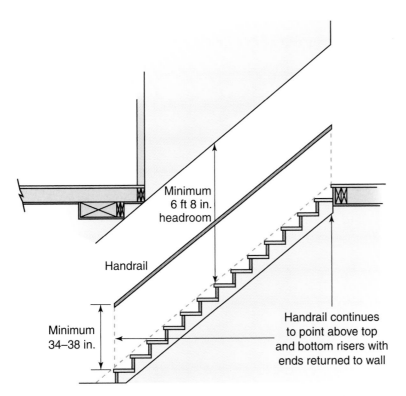

FIGURE 8-7 Stairway headroom and handrail height

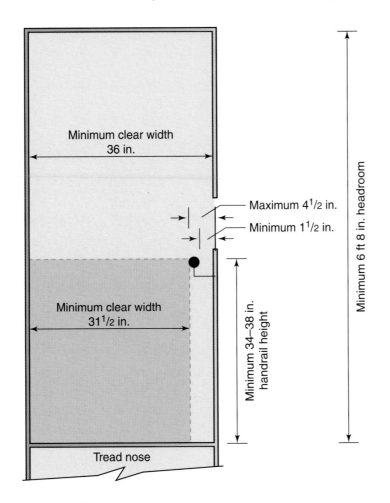

FIGURE 8-8 Stairway section, minimum width and height

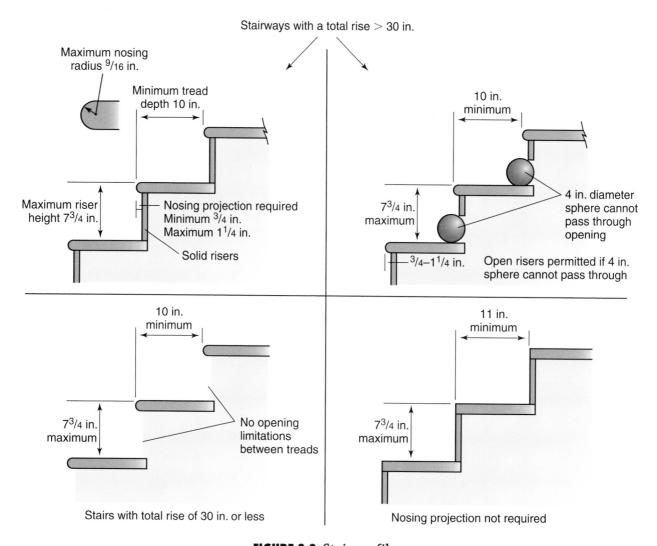

FIGURE 8-9 Stair profiles

distance from the walking surface to the grade or floor below is measured to the lowest point within 36 inches horizontally from the edge of the open sided walking surface. The 36-inch horizontal measurement accounts for the increased hazard of a steep slope or sudden drop-off near a deck or porch. The minimum guard height is usually measured from the walking surface, but for guards located adjacent to fixed seating, the minimum 36-inch height is measured above the seat and intends to protect children playing or climbing on the seat from falling over the guard (Figure 8-15).

At the sides of stairs, the minimum guard height is reduced to 34 inches to correlate with the minimum handrail height. The top rail of a stair guard often also serves as the stair handrail. Guards also must be constructed in such a way that a 4-inch sphere will not pass through, a dimension determined after lengthy research to prevent small children from maneuvering through such a barrier. The code grants two exceptions at the sides of stairs. The first increases the dimension to a 6-inch sphere at the triangle formed by the tread, riser, and bottom rail because of the impracticality of reducing the triangle and the negligible hazard. The second stipulates that a 4⅜-inch sphere cannot pass through a guard on the sides of stairs, a measurement that accommodates a practical

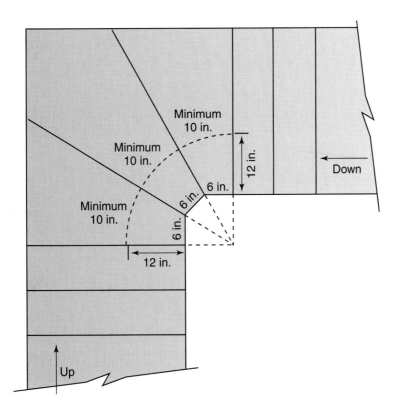

FIGURE 8-10 Winders

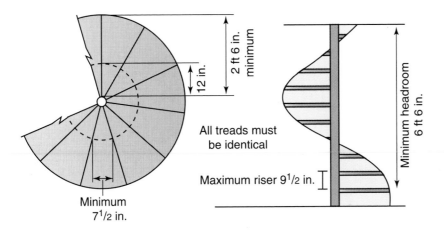

FIGURE 8-11 Spiral stair

wood spindle layout for staircases. Both exceptions are reasonable compromises, particularly when considered against the greater hazard of a child falling down the stair itself (Figure 8-16).

The height and allowable openings in guards are prescriptive requirements that are objectively measurable for compliance with the code. On the other hand, construction of a top rail to resist a single concentrated load of 200 pounds applied in any direction and the infill components to resist a 50-pound horizontal load applied to an area of 1 square foot are performance requirements. These requirements are not so easily measured or verified, though an experienced builder or inspector may fairly accurately gage the stiffness of a guard or handrail assembly.

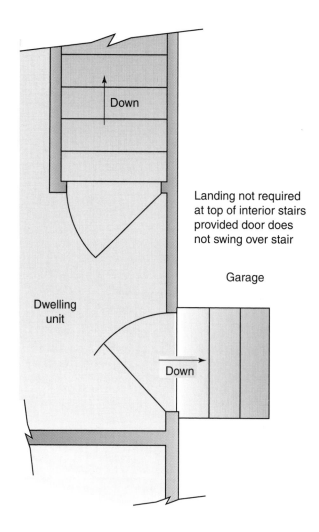

Landing not required
at top of interior stairs
provided door does
not swing over stair

Garage

Dwelling
unit

Down

Down

FIGURE 8-12 Door at interior stairways

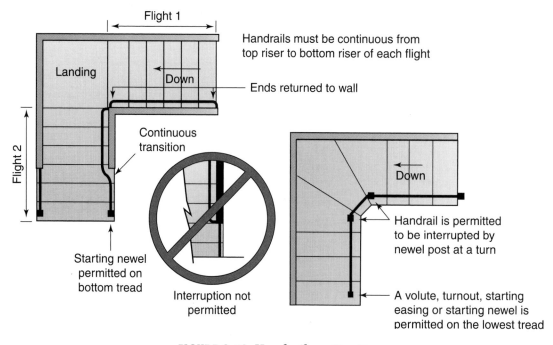

Flight 1

Landing

Down

Flight 2

Handrails must be continuous from
top riser to bottom riser of each flight

Ends returned to wall

Continuous
transition

Starting newel
permitted on
bottom tread

Interruption not
permitted

Down

Handrail is permitted
to be interrupted by
newel post at a turn

A volute, turnout, starting
easing or starting newel is
permitted on the lowest tread

FIGURE 8-13 Handrail continuity

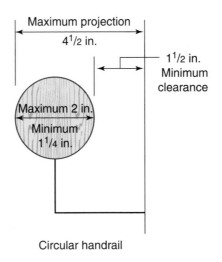

Circular handrail

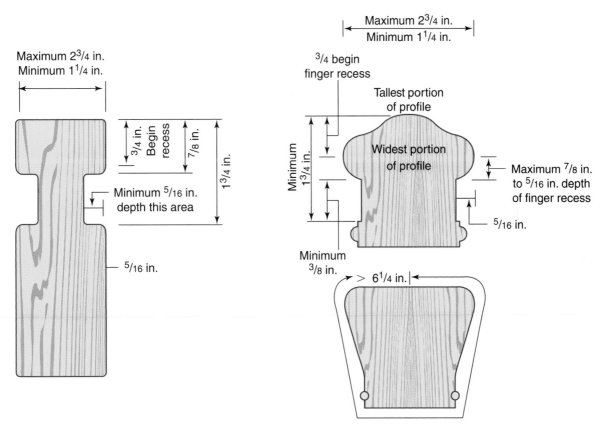

Type II handrails
perimeter > 6¹/₄ in.

FIGURE 8-14 Handrail shapes for graspability

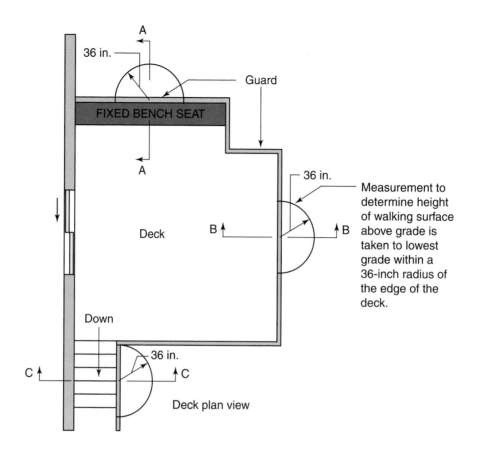

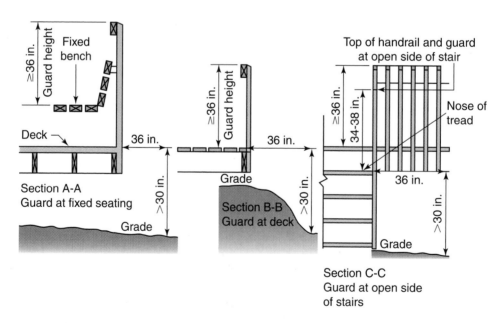

FIGURE 8-15 Determining guard locations and minimum heights

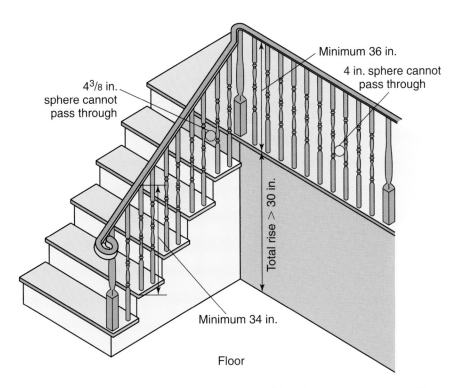

Minimum 36 in.

4 in. sphere cannot pass through

4³/₈ in. sphere cannot pass through

Total rise > 30 in.

Minimum 34 in.

Floor

FIGURE 8-16 Guard at interior stair and landing open on one side

While the most common type of guard may be constructed of wood or steel balusters and a top rail, the code by no means limits the materials or methods of construction when the guard performs to the stated criteria. [Ref. R312]

Window-sill height

The minimum window-sill height requirements are intended to reduce the number of injuries from falls by children through open windows. The 24-inch sill height is typically above a small child's center of gravity, reducing the likelihood of the child's toppling over the sill. The code regulates this minimum sill height only when the window opening is more than 72 inches above the grade below. In such locations, where the sill height is lower than 24 inches, protection can be achieved by installing a barrier or limiting the dimensions of the window opening. One such alternative is a *window fall protection device* meeting the requirements of the referenced standard, thereby complying with the operation provisions for emergency escape and rescue openings. An approved *window opening limiting device* with a clearly labeled quick release device may also be used on emergency escape and rescue openings (Figures 8-17 and 8-18). [Ref. R612.2]

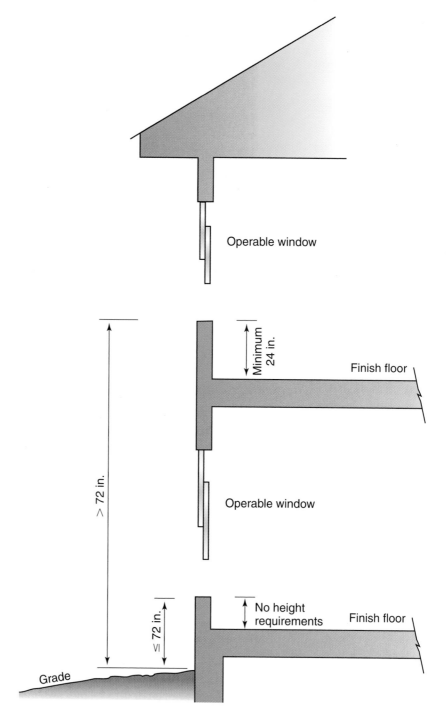

FIGURE 8-17 Window-sill height

EMERGENCY ESCAPE AND RESCUE OPENINGS

One of the most important safety provisions in the IRC concerns openings for emergency escapes and rescues. These openings provide alternate means to escape from a sleeping room or basement in the event that a fire or other emergency blocks the usual path of egress. They allow occupants to escape directly to the safety of the outdoors and

Window opening limiting devices and fall prevention devices must be approved for emergency escape and rescue provisions

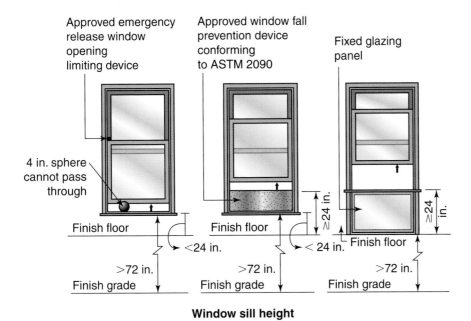

Window sill height

FIGURE 8-18 Alternatives to 24-inch window-sill height

allow rescue personnel fully equipped with breathing apparatus to enter the room from the outside. Occupants are most vulnerable to the hazards of fire when they are not fully alert or when they are occupying a basement, a space that traditionally has few windows or doors and often serves as a play or recreation area. The code addresses these life-safety issues by requiring an emergency escape and rescue opening in the basement and in every sleeping room. In addition, habitable attics require an emergency escape and rescue opening.

In order for emergency escape and rescue openings to effectively serve their intended purpose, the code prescribes a maximum sill height above the floor of 44 inches and a minimum net opening size of 5.7 square feet (5.0 square feet if the sill is not more than 44 inches above or below the finish grade). Width and height may be any number of combinations to achieve the minimum required opening area, provided the net width of the opening is not less than 20 inches and the net height not less than 24 inches (Table 8-1 and Figures 8-19 and 8-20). **[Ref. R310]**

In an emergency, occupants need to move quickly and easily to an outside space. Therefore, the code requires that the prescribed opening dimensions be obtained by the normal operation of the emergency escape and rescue opening, usually a window or door, without the need for a key, tool, or any special knowledge. This precludes the removal of a window sash or mechanical fasteners to obtain the required opening dimensions.

TABLE 8-1 Emergency escape and rescue openings*

		Inches				
Width	20	20.5	21	21.5	22	22.5
Height	41	40	39.1	38.2	37.3	36.5
Width	23	23.5	24	24.5	25	25.5
Height	35.7	34.9	34.2	33.5	32.8	32.2
Width	26	26.5	27	27.5	28	28.5
Height	31.6	31	30.4	29.8	29.3	28.8
Width	29	29.5	30	30.5	31	31.5
Height	28.3	27.8	27.4	26.9	26.5	26.1
Width	32	32.5	33	33.5	34	34.2
Height	25.7	25.3	24.9	24.5	24.1	24

*Minimum net clear width/height combinations to obtain a net opening of 5.7 square feet.

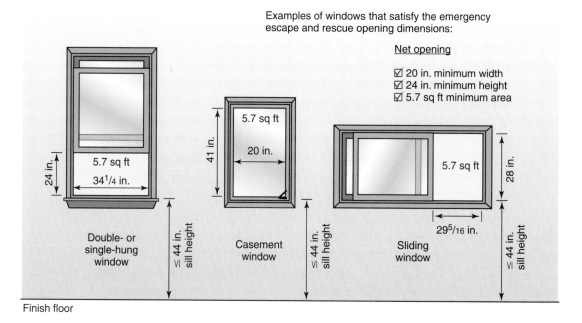

FIGURE 8-19 Emergency escape and rescue windows

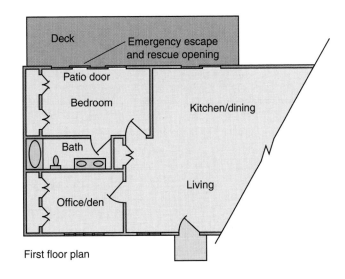

First floor plan

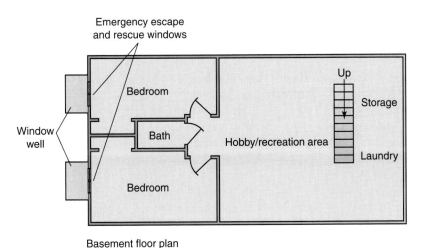

Basement floor plan

FIGURE 8-20 Required locations for emergency escape and rescue openings

Window wells

The IRC requires a window well when the sill of the emergency escape and rescue opening is below the adjacent ground elevation. The window well area must be at least 9 square feet with a minimum dimension of 36 inches. The code requires a ladder or steps when the window well is greater than 44 inches deep, a dimension consistent with the maximum window-sill height. A ladder is allowed to encroach into the required window well area no more than 6 inches (Figures 8-21 and 8-22). [**Ref. R310**]

SAFETY GLASS

To prevent serious injury from shards of breaking glass, the IRC requires safety glazing at eight specific glazing locations identified as subject to impact by people. For example, glass in doors and adjacent to doors has an increased likelihood of accidental breakage due to actions to open and

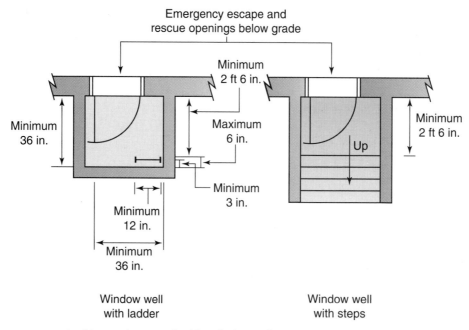

Emergency escape and
rescue openings below grade

Minimum
2 ft 6 in.

Minimum
36 in.

Maximum
6 in.

Minimum
3 in.

Minimum
12 in.

Minimum
36 in.

Up

Minimum
2 ft 6 in.

Window well
with ladder

Window well
with steps

Ladder or steps required for window wells greater than 44 in. below grade

FIGURE 8-21 Window wells for emergency escape and rescue windows

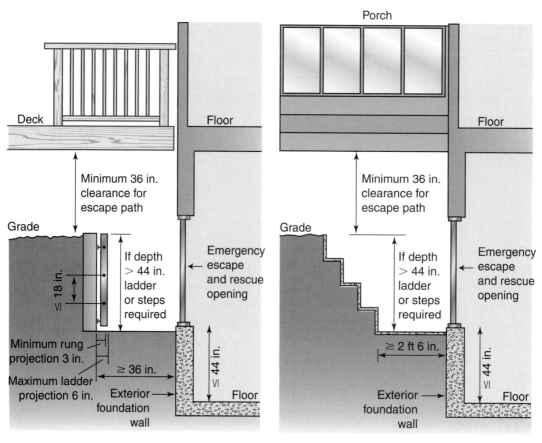

Deck

Floor

Porch

Floor

Minimum 36 in.
clearance for
escape path

Minimum 36 in.
clearance for
escape path

Grade

≤ 18 in.

If depth
> 44 in.
ladder
or steps
required

Emergency
escape
and rescue
opening

Grade

If depth
> 44 in.
ladder
or steps
required

Emergency
escape
and rescue
opening

Minimum rung
projection 3 in.

Maximum ladder
projection 6 in.

≥ 36 in.

Exterior
foundation
wall

44 in.

Floor

≥ 2 ft 6 in.

Exterior
foundation
wall

44 in.

Floor

Section at window well with ladder

Section at window well with ladder

FIGURE 8-22 Window well section views

close the door and the movement of the door itself. Large panels of glass lack the visual cues or physical barriers to prevent people from accidentally walking into them. Glass adjacent to stairs is considered hazardous because of the increased chances of a misstep or fall. Examples of hazardous locations subject to human impact and requiring safety glazing are illustrated in Figures 8-23 through 8-25. [Ref. R308]

Safety glazing, typically tempered or laminated glass, must pass the test requirements and be classified in accordance with the applicable referenced standard based on the location of the glazing. Polished wired glass is not permitted in hazardous locations requiring safety glazing, including those locations requiring fire resistance, unless it has received an approved classification through testing. The code generally requires a permanent manufacturer's designation on each panel of safety glazing indicating the type of glazing and the applicable standard. The designation is typically etched or embossed on the glass, but a label may be used, provided that it cannot be removed without being destroyed, thus preventing its reuse on a panel that is not safety glazing. For other than tempered glass, the building official may approve a certificate or affidavit of code compliance in lieu of a manufacturer's designation. [Ref. R308.1 and R308.3]

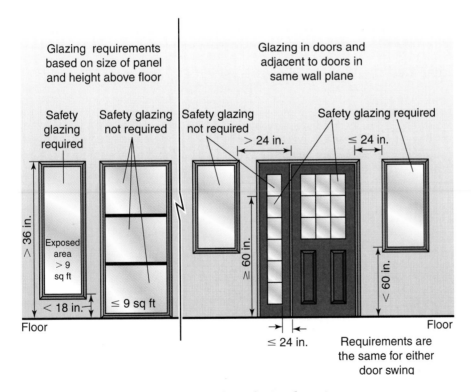

FIGURE 8-23 Safety glazing locations

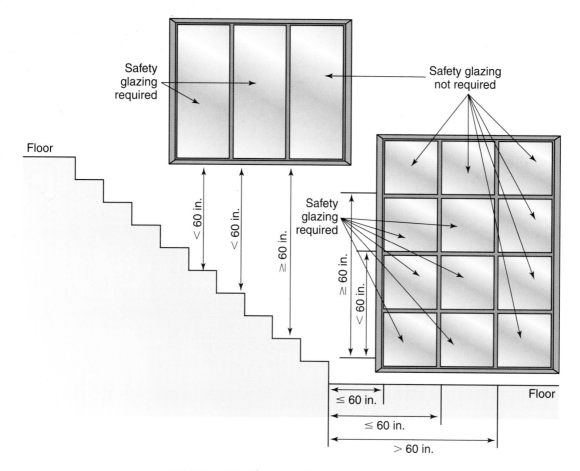

FIGURE 8-24 Glazing adjacent stairways

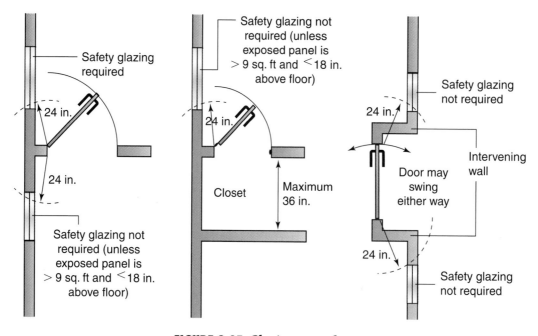

FIGURE 8-25 Glazing near doors

Fire Safety

Construction under the *International Residential Code* (IRC) protects occupants from the hazards of fire through interconnected smoke alarms, an automatic fire sprinkler system, and installation of fire resistant materials at prescribed locations (Figure 9-1). Such fire protection systems are concerned primarily with life safety and provide occupants time to safely exit the building. Although smoke alarms, residential sprinkler systems, and fire resistant construction often buy time for effective fire fighting, resulting in reduced property damage, such protection is secondary to occupant safety.

FIGURE 9-1 Fire sprinkler

SMOKE ALARMS

Occupants are most vulnerable to the hazards of fire while sleeping. Detection and notification in the early stages of a fire provide residents with needed time to escape before the interior environment becomes intolerable. The IRC requires a smoke alarm in each sleeping room, outside each sleeping area, and on each additional story of the dwelling unit including basements and habitable attics (Figure 9-2). The code also stipulates that the building wiring system provide the primary power to the smoke alarms and that batteries supply backup power when primary power is interrupted. [Ref. R314]

A smoke alarm is a self-contained device that provides both smoke detection and an alarm-sounding appliance. Smoke alarms must be listed as conforming to UL 217, *Single and Multiple Station Smoke Alarms*. The IRC requires interconnection of the devices such that when smoke is detected in one location, the alarms are activated in all locations. In addition to alerting residents on any story of the dwelling unit, interconnection ensures that the alarm is delivered to each bedroom at a sound-pressure level considered sufficient to wake a sleeping person. [Ref. R314.1, R314.3, and R314.4]

As an alternative to smoke alarms, the IRC permits a fire alarm system installed in accordance with the household fire warning equipment provisions of the National Fire Protection Association in NFPA 72, *National Fire Alarm Code*. A household fire alarm system typically has separate devices for smoke detection and alarm annunciation. Any such system must provide detection and notification equivalent to the IRC-prescribed requirements for smoke alarms. For example, a household fire warning system may have fewer notification devices placed in the building. When a detector in any of the prescribed locations activates the system, the alarm must be clearly audible in all bedrooms over background noise levels with all intervening doors closed. In general, the sound-pressure level at the pillow cannot be less than 70 decibels. [Ref. R314.2]

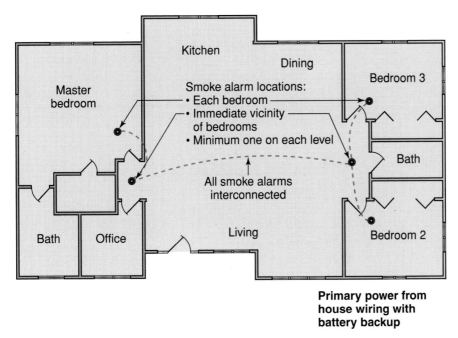

FIGURE 9-2 Smoke alarm locations

Smoke alarms in existing dwellings

The IRC also regulates smoke alarms for existing dwellings when interior alterations or repairs requiring a building permit occur or when an addition other than a deck or porch is constructed. In these cases, the building must be brought into conformance with the smoke alarm requirements for new buildings. If there are practical difficulties in installing wiring to the devices without damaging existing finishes, the code allows battery-operated smoke alarms without interconnection. The smoke alarm provisions do not apply in the case of minor work that does not require a permit, for exterior renovations, or for the addition of a deck or porch. **[Ref. R314.3.1]**

FIRE SPRINKLER SYSTEMS

An automatic fire sprinkler system conforming to IRC Section P2904, *Dwelling Unit Fire Sprinkler Systems*, or NFPA 13D, *Installation of Sprinkler Systems in One- and Two-Family Dwellings and Manufactured Homes*, aids in the detection and control of fires in dwellings and intends to prevent total fire involvement (flashover) in the room of fire origin for a period of time to allow the escape or evacuation of the dwelling occupants (Figure 9-3). The IRC requires an automatic fire sprinkler system installed in accordance with Section P2904 or NFPA 13D in all new one- and two-family dwellings and townhouses. The townhouse requirements are effective upon adoption of the IRC, while the one- and two-family dwelling provisions become effective January 1, 2011.

IRC Section P2904 provides a simple, prescriptive approach to the design of dwelling fire sprinkler systems and is an approved alternative to NFPA 13D, which allows for engineered design options and other piping configurations. Consistent with structural and other design

FIGURE 9-3 Residential sprinkler rough-in using approved CPVC pipe

provisions in the code, the IRC prescriptive methods allow contactors, plumbers, and homeowners to design and install a dwelling sprinkler system while still providing the flexibility of an engineered design in accordance with NFPA 13D.

Although the title of NFPA 13D suggests that the application of the standard is limited to one- and two-family dwellings, it is also appropriate for automatic sprinkler systems installed in townhouses regulated by the IRC. Each dwelling unit in a townhouse has a separate connection to water, wastewater, fuel-gas, and electrical utilities, and such connections are under the control of the dwelling unit occupant. Conversely, an NFPA 13R system is designed for the protection of apartment buildings, lodging and rooming houses, and hotels, motels, and dormitories up to four stories in height. In these occupancies, the property management is responsible for the maintenance of the entire building in accordance with the IBC, the *International Fire Code* (IFC), and the *International Property Maintenance Code* (IPMC).

A dwelling fire sprinkler system requires less water when compared to NFPA 13 and 13R systems. Section P2904 and NFPA 13D generally require a minimum water discharge duration of 10 minutes, compared to 30 minutes for an NFPA 13R system and even higher duration values in NFPA 13. The minimum sprinkler discharge density may be satisfied by connection to a domestic water supply, a water well, an elevated storage tank, an approved pressure tank, or a stored water source with an automatically operated pump. Any combination of water supply systems is allowed to meet the required dwelling fire sprinkler system capacity.

Unlike an NFPA 13 automatic sprinkler system, a dwelling fire sprinkler system does not require automatic sprinkler protection throughout a one- and two-family dwelling or townhouse. Sprinklers are not required in areas that have been statistically shown through fire incident loss data to not significantly contribute to injuries or death. For example, sprinklers may be omitted in closets with areas of 24 square feet or less and bathrooms with areas of 55 square feet or less. Generally, sprinklers also are not required in open attached porches, garages, attics, crawl spaces, and concealed spaces not intended or used for living purposes. In addition, a dwelling fire sprinkler system does not require a fire department connection.

The provisions for dwelling fire sprinkler systems require the use of new sprinklers listed for residential applications. These sprinklers have been investigated and listed for use inside of dwelling units. They are equipped with a thermal element using either a fusible link or frangible bulb filled with conductive liquid that is designed to operate approximately five times faster when compared to a standard spray sprinkler required by NFPA 13 installed in the same setting. The sprinkler discharge pattern is designed to wet the walls and floors of the space and apply water onto any furnishings. [Ref. R313 and P2904]

SEPARATION BETWEEN DWELLING UNITS

Fire resistance rated construction intends to confine a fire to a given area for a period of time. The components of the wall or floor-ceiling construction form an assembly that has proven through testing to resist the effects of fire for the designated time period. The fire resistance rating is based on results under the specific test conditions and does not necessarily predict performance under actual field conditions.

Occupants have no control over the action of their neighbors, and the fire resistant separation between dwelling units offers an appropriate level of protection.

Two-family dwellings

The IRC requires a one-hour fire resistance rated separation between the dwelling units of a two-family dwelling. Horizontal floor-ceiling assemblies separating upper and lower units must extend to the exterior walls, and supporting wall construction must also be one-hour fire resistance rated (Figure 9-4). Wall assemblies separating side-by-side units must generally extend from the foundation through the attic space to the bottom of the roof sheathing (Figure 9-5). As an alternative, the IRC permits the wall assembly to terminate at a ⅝-inch gypsum board ceiling when a draft stop is installed in the attic area and not less than ½-inch gypsum board is installed on the walls supporting the ceiling (Figure 9-6). The code allows a reduction in the fire resistance rated separation to a half hour when an NFPA 13 automatic fire sprinkler system is installed throughout the building. The code does not permit an NFPA 13R or 13D residential system for this trade-off. Automatic sprinkler systems in conformance with NFPA 13 require complete sprinkler protection throughout a building, including attics and concealed floor-ceiling areas, and offer a higher level of protection to the building. NFPA 13–compliant systems are approved for installation in any type of occupancy. [Ref. R302.3]

Townhouses

The permissable number of townhouses in a building is unlimited, and the IRC treats their separation somewhat differently. Each townhouse is considered a separate building for separation purposes, with the joining

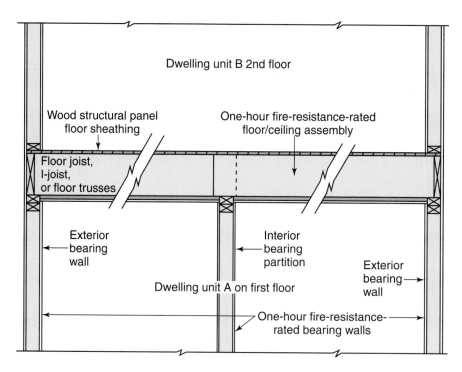

FIGURE 9-4 Horizontal separation for two-family dwelling

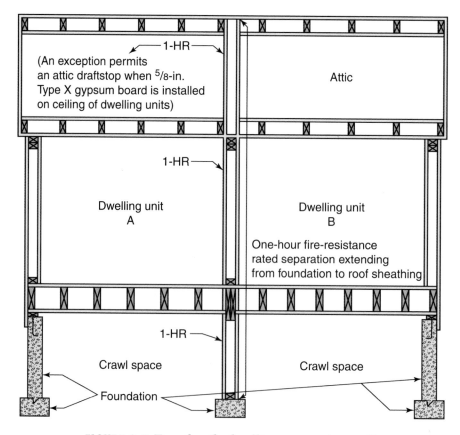

FIGURE 9-5 Two-family dwelling separation wall

walls of each unit treated as exterior walls on a property line. Therefore, each townhouse requires a one-hour fire resistance rated wall at the separation, and each townhouse must generally be structurally independent. In other words, structural elements, such as roof framing, do not depend

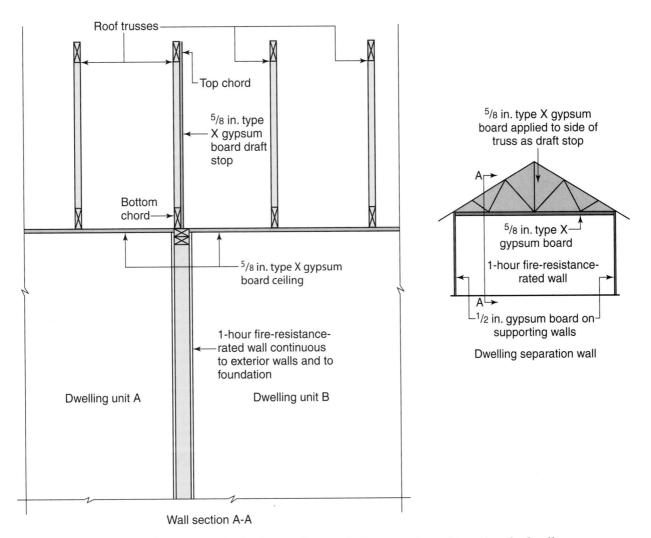

FIGURE 9-6 Alternate attic draft stop for vertical separation of two-family dwelling

on adjoining townhouse construction for their support. However, there are some practical difficulties in constructing separate one-hour-rated walls, and the IRC allows a common one-hour fire resistance rated separation wall that is exempt from the structural independence requirement. For example, this common wall is permitted to support floor and roof framing elements of both townhouses. The code prohibits the installation of any plumbing or mechanical work in the common wall cavity. In most cases, builders will opt for the one-hour common wall as the preferred method of construction (Figure 9-7). **[Ref. R302.1 and R302.2]**

To prevent the spread of fire from one unit to the next at the roof line, the IRC provides two options. The first requires a 30-inch-high parapet wall with equivalent fire rating as the separation wall. The second option is the more common practice of installing fire resistant protection at the roof for a distance of 4 feet on each side of the separating wall (Figures 9-8 and 9-9).

Fire resistance rated assemblies

Many tested fire resistance rated assemblies are available utilizing various materials and methods of construction. In buildings regulated by the IRC, gypsum board applied to wood framing is the most common type of

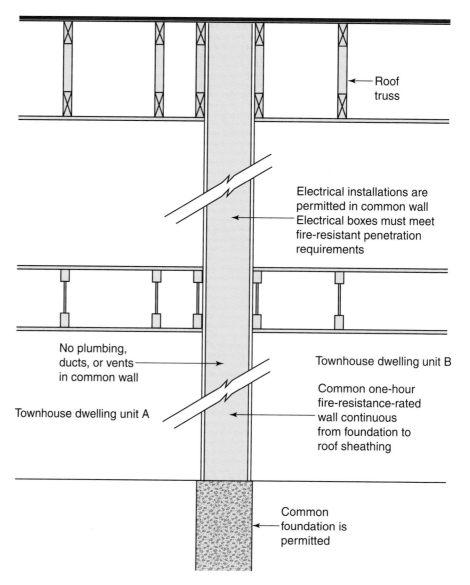

Roof truss

Electrical installations are permitted in common wall Electrical boxes must meet fire-resistant penetration requirements

No plumbing, ducts, or vents in common wall

Townhouse dwelling unit B

Townhouse dwelling unit A

Common one-hour fire-resistance-rated wall continuous from foundation to roof sheathing

Common foundation is permitted

FIGURE 9-7 Common one-hour fire resistance rated wall for townhouse separation

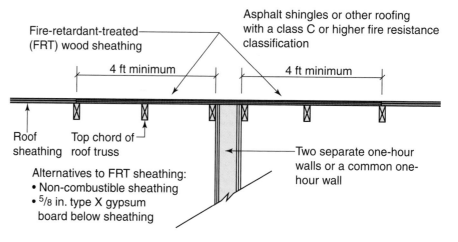

Fire-retardant-treated (FRT) wood sheathing

Asphalt shingles or other roofing with a class C or higher fire resistance classification

4 ft minimum

4 ft minimum

Roof sheathing

Top chord of roof truss

Alternatives to FRT sheathing:
• Non-combustible sheathing
• $5/8$ in. type X gypsum board below sheathing

Two separate one-hour walls or a common one-hour wall

FIGURE 9-8 Townhouse roof protection on each side of separation wall

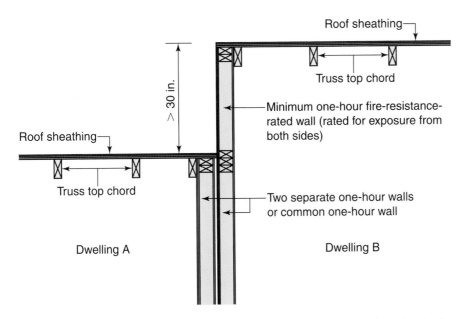

FIGURE 9-9 Townhouse separation for roofs with greater than 30-inch height difference

assembly. Assemblies are assigned an hourly fire resistance rating through testing in accordance with American Society for Testing and Materials standard ASTM E 119, *Test Methods for Fire Tests of Building Materials and Construction* or Underwriters Laboratories UL 263 *Fire Tests of Building Construction and Materials*. Tested assemblies are available in the *Gypsum Association Fire Resistance Design Manual* and from approved testing agencies. Construction must match the design specifications for types of materials, dimensions, and methods of attachment (Figures 9-10 and 9-11). **[Ref. R302.1, R302.2, and R302.3]**

Penetrations of fire resistance rated assemblies

When items such as pipes or ducts penetrate one or both sides of the fire resistance rated wall or floor-ceiling assembly separating dwelling units, both the penetrating item and the space around it must be protected to maintain the integrity of the fire resistant assembly. In general, penetrations by metal pipe require that the space around the pipe be filled with approved materials to prevent the passage of flame and hot gases. The material, such as a listed fire-stop sealant, must be installed according to the manufacturer's instructions to provide a fire resistant time rating that is equivalent to that of the construction being penetrated. Other penetrating materials, such as plastic pipe (if permissible—plumbing piping is not permitted to penetrate or be located in a one-hour common wall separating townhouses), must be protected by an approved penetration fire-stop system. Such a system is often a collar containing intumescent material that expands when heated by fire conditions, filling the penetration as the plastic pipe melts and preserving the fire resistance rating of the wall or floor-ceiling assembly. **[Ref. R302.4]**

Steel electrical boxes up to 16 square inches and listed electrical boxes are permitted to penetrate the membrane of a fire resistance rated wall assembly. Steel box penetrations are limited to 100 square inches in any 100 square feet of wall area. When electrical boxes are installed on

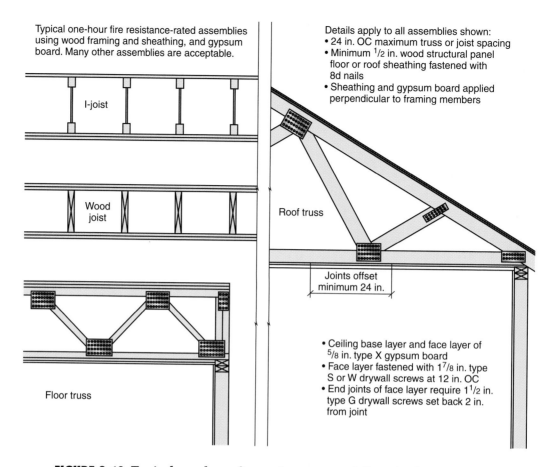

Typical one-hour fire resistance-rated assemblies using wood framing and sheathing, and gypsum board. Many other assemblies are acceptable.

I-joist

Wood joist

Floor truss

Details apply to all assemblies shown:
• 24 in. OC maximum truss or joist spacing
• Minimum ¹/₂ in. wood structural panel floor or roof sheathing fastened with 8d nails
• Sheathing and gypsum board applied perpendicular to framing members

Roof truss

Joints offset minimum 24 in.

• Ceiling base layer and face layer of ⁵/₈ in. type X gypsum board
• Face layer fastened with 1⁷/₈ in. type S or W drywall screws at 12 in. OC
• End joints of face layer require 1¹/₂ in. type G drywall screws set back 2 in. from joint

FIGURE 9-10 Typical one-hour fire resistance rated floor/ceiling and roof/ceiling assemblies

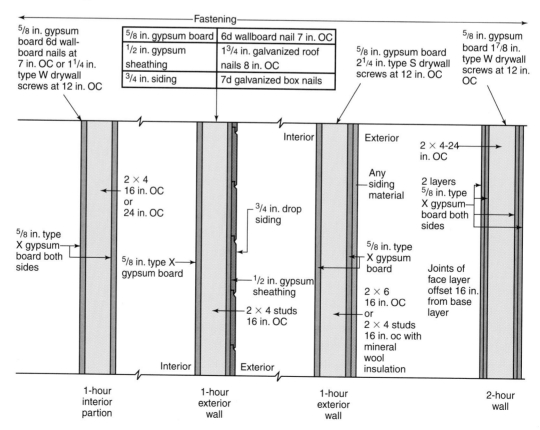

Fastening

⁵/₈ in. gypsum board 6d wall-board nails at 7 in. OC or 1¹/₄ in. type W drywall screws at 12 in. OC

⁵/₈ in. gypsum board	6d wallboard nail 7 in. OC
¹/₂ in. gypsum sheathing	1³/₄ in. galvanized roof nails 8 in. OC
³/₄ in. siding	7d galvanized box nails

⁵/₈ in. gypsum board 2¹/₄ in. type S drywall screws at 12 in. OC

⁵/₈ in. gypsum board 1⁷/₈ in. type W drywall screws at 12 in. OC

Interior   Exterior

2 × 4-24 in. OC

Any siding material

2 layers ⁵/₈ in. type X gypsum board both sides

⁵/₈ in. type X gypsum board both sides

2 × 4 16 in. OC or 24 in. OC

⁵/₈ in. type X gypsum board

³/₄ in. drop siding

¹/₂ in. gypsum sheathing

2 × 4 studs 16 in. OC

⁵/₈ in. type X gypsum board

2 × 6 16 in. OC or 2 × 4 studs 16 in. oc with mineral wool insulation

Joints of face layer offset 16 in. from base layer

Interior   Exterior

1-hour interior partition

1-hour exterior wall

1-hour exterior wall

2-hour wall

FIGURE 9-11 Typical fire resistance rated wall assemblies

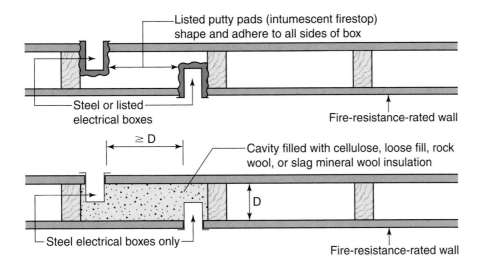

Note: Steel boxes limited to 16 sq. in. each and total of 100 sq. in. in any 100 sq. ft of wall

Listed putty pads (intumescent firestop) shape and adhere to all sides of box

Steel or listed electrical boxes

Fire-resistance-rated wall

≥ D

Cavity filled with cellulose, loose fill, rock wool, or slag mineral wool insulation

D

Steel electrical boxes only

Fire-resistance-rated wall

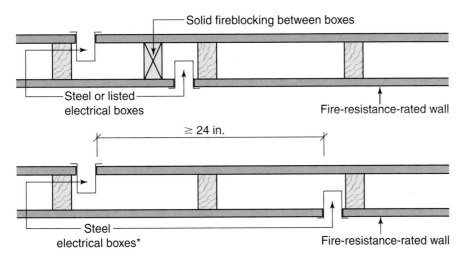

Solid fireblocking between boxes

Steel or listed electrical boxes

Fire-resistance-rated wall

≥ 24 in.

Steel electrical boxes*

Fire-resistance-rated wall

FIGURE 9-12 Electrical box penetrations on opposite sides of fire resistance rated wall assembly

*Note: Separation distance for listed boxes is determined by the listing.

opposite sides of the wall assembly, they require a minimum horizontal separation distance or another approved means of separation. The installation of wood blocking between the boxes or application of a listed putty pad to each box satisfies the separation requirement. Putty pads consist of intumescent fire-stop material that expands when heated to seal off the electrical box penetration (Figure 9-12). [**Ref. R302.4.2**]

DWELLING SEPARATION FROM GARAGE

Unlike separations between dwelling units, the separation between the residence and garage is not a fire resistance rated assembly. Likewise, penetrations through the separation are not required to meet the rated penetration requirements for fire resistance rated assemblies. Attached garages and detached garages within 3 feet of the dwelling require installation of gypsum board on the garage side to provide limited resistance to the spread of fire. Generally, the IRC prescribes ½-inch gypsum board installed on the garage side to achieve this separation (Figures 9-13 and 9-14). When there are habitable rooms above the garage, the code requires the installation of ⅝-inch Type X gypsum board on the garage ceiling. The bearing

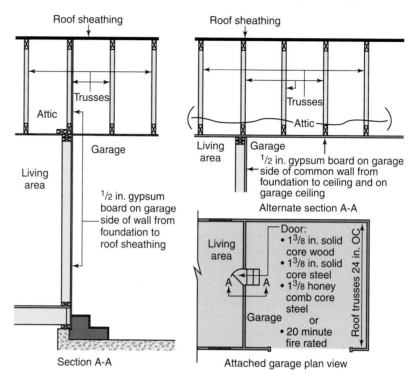

FIGURE 9-13 Dwelling and attached garage separation

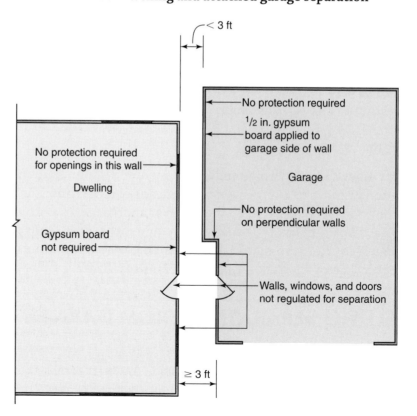

FIGURE 9-14 Dwelling and detached garage separation

walls supporting the ceiling framing in this instance also require the application of ½-inch gypsum board on the interior surface (Figure 9-15). **[Ref. R302.6]**

Doors between the dwelling and the garage also provide some resistance to fire but do not require an assembly with a fire resistance rating. In other words, the frame, hardware, and sealing of the opening are not addressed, only the materials of the door leaf itself. Any one of the following types of door satisfies the separation requirement:

- 1⅜-inch-thick solid-core wood
- 1⅜-inch-thick solid-core steel
- 1⅜-inch-thick honeycomb-core steel
- A listed door with a 20-minute fire resistance rating

Openings from the garage into a sleeping room are prohibited.

The IRC requires minimum No. 26-gauge sheet steel for ducts in the garage as well as ducts penetrating the walls or ceilings that separate the dwelling from the garage. Ducts are not permitted to open into the garage. Other penetrations, such as plastic or steel pipe, require only that the space around the penetration be filled with approved materials to limit the free passage of fire and smoke. **[Ref. R302.5]**

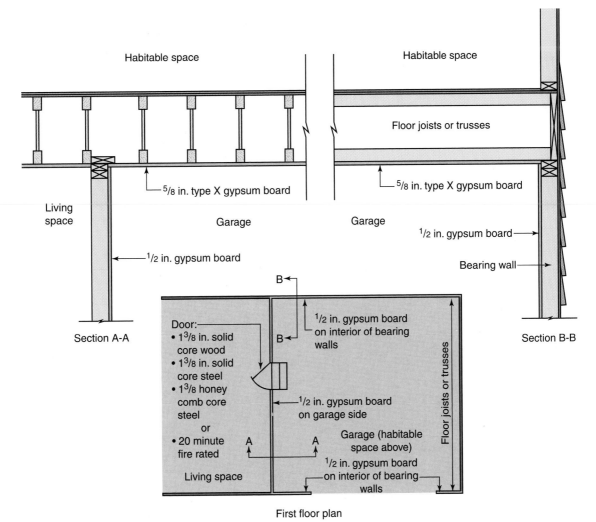

FIGURE 9-15 Separation for attached garage with habitable space above

FOAM PLASTIC

In addition to meeting the maximum flame spread and smoke-developed indices, foam plastic insulation must be isolated from exposure to fire from the dwelling interior by the installation of a thermal barrier. The IRC prescribes the application of ½-inch gypsum board or an approved material providing equivalent protection to separate the foam plastic from the interior (Figure 9-16). In attics and crawl spaces entered only

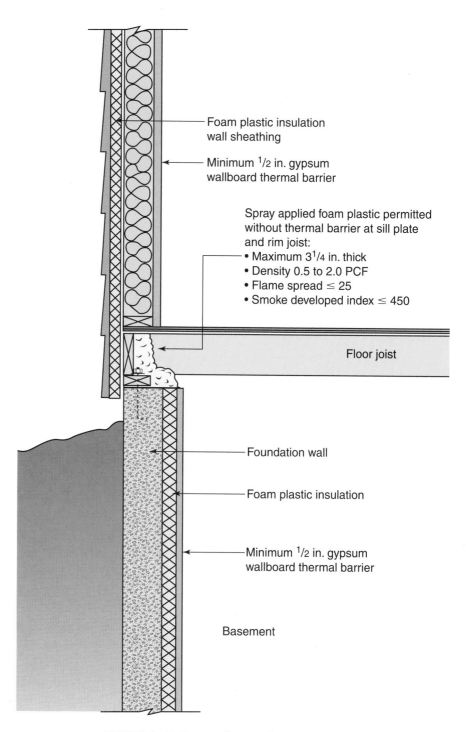

Foam plastic insulation wall sheathing

Minimum ¹/2 in. gypsum wallboard thermal barrier

Spray applied foam plastic permitted without thermal barrier at sill plate and rim joist:
• Maximum 3¹/4 in. thick
• Density 0.5 to 2.0 PCF
• Flame spread ≤ 25
• Smoke developed index ≤ 450

Floor joist

Foundation wall

Foam plastic insulation

Minimum ¹/2 in. gypsum wallboard thermal barrier

Basement

FIGURE 9-16 Foam plastic thermal barrier

for maintenance or repairs, the IRC permits a reduction in protection, and the foam plastic may be covered with one of the following ignition barrier materials (Figure 9-17):

- 1½-inch-thick mineral fiber insulation
- ¼-inch wood structural panels
- ⅜-inch particleboard
- ¼-inch hardboard
- ⅜-inch gypsum board
- 0.016-inch corrosion-resistant steel

[**Ref. R316.4 and R316.5**]

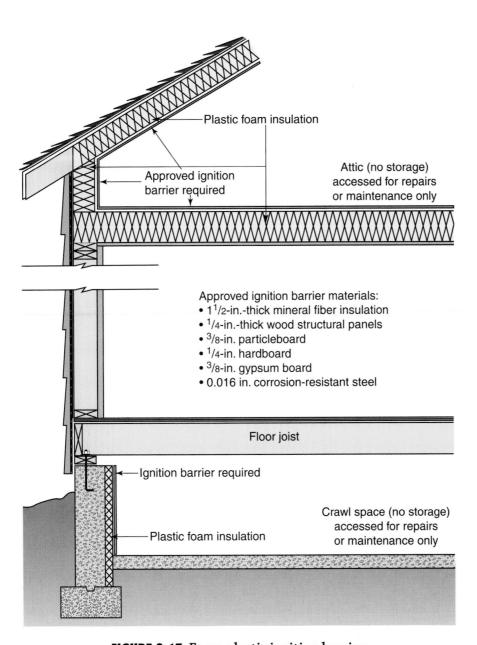

Plastic foam insulation

Approved ignition barrier required

Attic (no storage) accessed for repairs or maintenance only

Approved ignition barrier materials:
- 1$\frac{1}{2}$-in.-thick mineral fiber insulation
- $\frac{1}{4}$-in.-thick wood structural panels
- $\frac{3}{8}$-in. particleboard
- $\frac{1}{4}$-in. hardboard
- $\frac{3}{8}$-in. gypsum board
- 0.016 in. corrosion-resistant steel

Floor joist

Ignition barrier required

Plastic foam insulation

Crawl space (no storage) accessed for repairs or maintenance only

FIGURE 9-17 Foam plastic ignition barrier

Healthy Living Environment

The *International Residential Code* (IRC) sets minimum requirements for natural or artificial light, fresh air ventilation, carbon monoxide alarms, comfort heating, and sanitation to create a healthy and livable environment (Figure 10-1).

FIGURE 10-1 Shower and tub space

NATURAL AND ARTIFICIAL LIGHT

Though the code retains the traditional standards for natural light from windows, electric lighting satisfies the minimum illumination requirements for habitable rooms in almost all cases. The minimum average illumination level for artificial lighting in habitable rooms is 6 footcandles, far below typical indoor illumination levels and lighting industry recommendations of 50 footcandles or more. While windows may be eliminated for lighting purposes, they may still be required for emergency escape and rescue and fresh air ventilation purposes. [Ref. R303.1]

Stairway illumination

As part of the egress path and a component presenting increased hazards of fall injuries, stairway design and construction, including adequate illumination, is particularly important to safety in a dwelling. The IRC requires a minimum illumination level of 1 footcandle at treads and landings of interior stairs. Light sources must be placed in the immediate vicinity of each landing or directly over each flight of stairs. For other than continuous or automatic illumination (such as provided with motion sensors), interior stairways with six or more risers require a wall switch at each floor level.

Exterior stairs require a light source located near the top landing. For other than continuous or automatic illumination, the IRC requires control by a wall switch located inside the dwelling (Figure 10-2). [Ref. R303.6 and E3903.3]

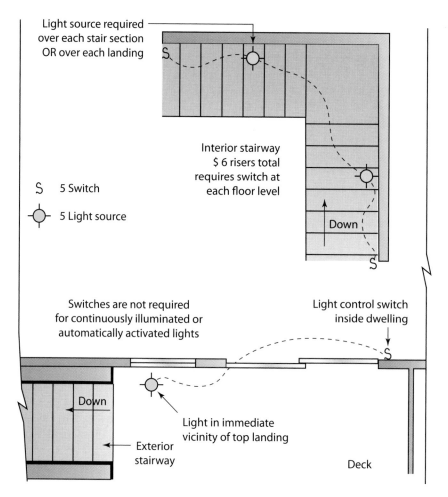

FIGURE 10-2 Stairway illumination

NATURAL AND MECHANICAL VENTILATION

To provide fresh air ventilation to habitable rooms, the IRC requires an openable window or door area to the outdoors equal to 4 percent of the floor area of the room or an approved mechanical ventilation system. Mechanical ventilation may be achieved in one of two ways. In the first, the system must be capable of providing 0.35 air changes per hour (ach) to each habitable room. Alternatively, a whole-house mechanical ventilation system may be installed to provide outside air at a rate of 15 cubic feet per minute per person. For purposes of whole-house ventilation, the number of occupants is based on the number of bedrooms, with two occupants in the first bedroom and one each in the remaining bedrooms. [Ref. R303.1]

Each bathroom requires a window(s) with 1.5 square feet of total openable area or mechanical ventilation exhausting directly to the outside air. The minimum mechanical ventilation rates are 50 cfm for intermittent operation and 20 cfm for continuous operation of the exhaust fan. [Ref. R303.3]

Openable windows, doors, mechanical ventilation air intakes, and similar openings that draw air from the outside must be located at least

> **EXAMPLE**
>
> ### Mechanical ventilation to habitable room
> Calculate the required ventilation in cubic feet per minute (cfm) for a habitable room measuring 12 by 16 feet with a ceiling height of 8 feet:
>
> $$12 \times 16 \times 8 = 1536 \text{ cubic feet.}$$
>
> The IRC requires 0.35 ach per hour. 1536 cubic feet $\times$ 0.35 = 537.6 cubic feet per hour (cfh). 537.6 cfh / 60 minutes = 8.96 cfm mechanical ventilation for this room.

> **EXAMPLE**
>
> ### Whole-house mechanical ventilation
> Calculate the whole-house ventilation rate for a four-bedroom house.
>
> | Bedrooms | Occupants |
> |---|---|
> | First | 2 |
> | Second | 1 |
> | Third | 1 |
> | Fourth | 1 |
> | Total | 5 |
>
> 5×15 cfm = 75 cfm whole-house outdoor air ventilation rate.

10 feet from any noxious source such as plumbing vents, flue vents, and chimneys. Openings located at least 2 feet below the noxious source are not subject to the 10-foot separation requirement. Exhaust and intake openings must be protected on the outside with corrosion-resistant screens with openings of ¼ to ½ inch (Figure 10-3). **[Ref. R303.4 and R303.5]**

CARBON MONOXIDE ALARMS

As part of a safe and healthy interior living environment, the IRC provides for early warning to alert occupants to hazardous levels of carbon monoxide gas. The code requires carbon monoxide alarms in new dwelling units and in existing dwelling units undergoing renovation requiring a permit. This installation requirement triggered by construction work on the existing dwelling is similar to the smoke alarm provisions. Unlike the smoke alarm requirements, there is no exception for exterior work or the addition of decks or porches. Roofing, siding, window replacement, and other exterior work requiring a permit requires the installation of carbon monoxide alarms.

Because the source of unsafe levels of carbon monoxide in the home is typically from faulty operation of a fuel-fired furnace or water heater, or from the exhaust of an automobile, this requirement only applies to homes containing fuel-fired appliances or having an attached garage. Carbon monoxide accumulates in the body over time relative to its

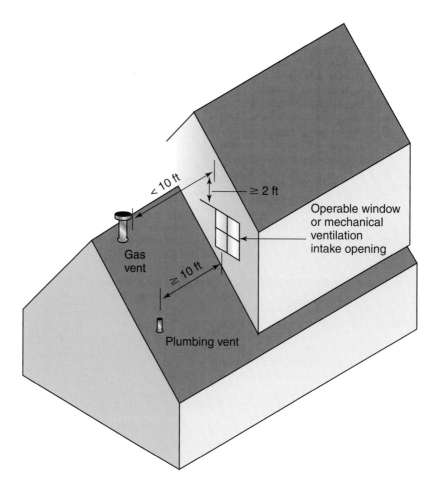

FIGURE 10-3 Outside air intake opening locations

concentration in the air. Accordingly, carbon monoxide detectors sound an alarm based on the concentration of carbon monoxide and the amount of time that certain levels are detected, simulating an accumulation of the toxic gas in the body. High levels of carbon monoxide will trigger an alarm within a short time, while lower levels must be present over a longer time period for the alarm to sound. This design prevents false-positive alarms. This code requirement recognizes the reliability of carbon monoxide alarms and the referenced standard, UL 2034, *Single and Multiple Station Carbon Monoxide Alarms*, and intends to reduce accidental deaths from carbon monoxide poisoning.

Because carbon monoxide poisoning deaths often occur when the occupant is sleeping, the IRC requires carbon monoxide alarms to be located in the areas outside of and adjacent to bedrooms (Figure 10-4). **[Ref. R315]**

HEATING AND COOLING

In geographical areas where the design temperature is less than 60°F, as indicated by the governing authority in the adopting ordinance (see Chapter 4 of this publication), the IRC requires a heating system capable of maintaining a minimum temperature of 68°F at 3 feet above the floor

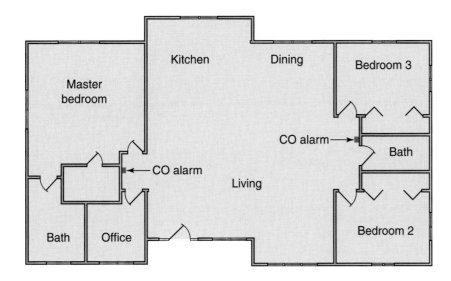

FIGURE 10-4 Carbon monoxide (CO) alarm installed in the immediate vicinity of each sleeping area

and 2 feet from exterior walls. The code does not require the installation of an air conditioning or comfort cooling system. When mechanical equipment for heating or cooling is installed, it must comply with the mechanical and fuel-gas provisions of the IRC (see Chapter 12 of this publication). **[Ref. R303.8]**

SANITATION

In the building planning chapter of the code, the IRC establishes basic requirements for bathroom and kitchen fixtures, clearance dimensions, hot and cold water, and sewer connection. Installation must also comply with the specific requirements of the IRC plumbing provisions (see Chapter 13 of this publication).

Toilet and bathing facilities

In order to maintain a healthy and sanitary living environment, a residence must provide facilities for toilet, bathing, and handwashing purposes. The IRC requires at least one water closet, one lavatory, and a bathtub or shower in every dwelling unit. Each fixture must be connected to an approved water supply and sewer. Lavatories, bathtubs, showers, and bidets require connection to both hot and cold water supply. **[Ref. R306]**

The IRC prescribes minimum clearance dimensions around bathroom fixtures so that occupants can reasonably access and use the fixtures. The minimum size of a shower is also set at 30 inches by 30 inches, though the IRC plumbing provisions provide an alternative for a narrower shower compartment with a greater area. In this case, the minimum inside width of the shower compartment is 25 inches, and the minimum inside area is 1300 square inches, which correlates to the approximate inside dimensions of a standard bathtub. These alternative dimensions are useful in remodeling situations where a stand-up shower replaces a bathtub (Figures 10-5 and 10-6). **[Ref. 307.1, P2705.1, and P2708.1]**

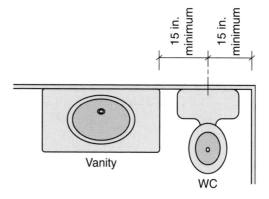

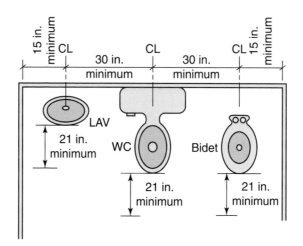

Installation of fixtures

FIGURE 10-5 Bathroom fixture clearances

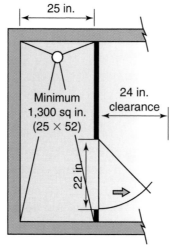

Alternate minimum shower dimensions (approximate finish dimensions of bathtub)

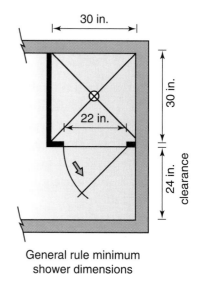

General rule minimum shower dimensions

FIGURE 10-6 Minimum shower dimensions

For ease of cleaning and to protect the structure from deterioration, showers must have nonabsorbent wall surfaces to a minimum height of 6 feet above the floor. [Ref. 307.2]

Cooking and cleaning facilities

Also important to the health of the occupants are adequate and sanitary means to prepare meals, wash dishes, and wash clothes. The IRC requires a kitchen area and kitchen sink for every dwelling unit. Kitchen sinks, laundry tubs, and washing machine outlets require connection to hot and cold water supply. [Ref. R306.2 and R306.4]

Code Basics

A dwelling unit is considered a single unit providing complete independent living facilities for one or more persons, including permanent provisions for:

- Living
- Sleeping
- Eating
- Cooking
- Sanitation

Chimneys and Fireplaces

The *International Residential Code* (IRC) contains prescriptive provisions for the construction of masonry chimneys and fireplaces, including requirements for combustion air supply, clearance to combustibles, and hearth construction (Figure 11-1). The appliance listing and manufacturer's instructions typically govern the installation of approved factory-built fireplaces and chimneys.

FIGURE 11-1 A masonry chimney is constructed of solid masonry units, hollow masonry units grouted solid, stone, or concrete

EXTERIOR AIR SUPPLY

Both factory-built and masonry fireplaces require an adequate exterior air supply to assure proper fuel combustion and prevent depleting oxygen within the habitable space. Mechanical ventilation of the room is permitted as an alternative when controlled so that the indoor pressure of the room or space is neutral or positive. Exterior combustion air ducts or passageways for masonry fireplaces must be at least 6 square inches and not more than 55 square inches in cross-section area. Listed ducts for masonry fireplaces must be installed according to the terms of their listing and the manufacturer's instructions. Factory-built fireplaces require exterior air ducts that are a listed component of the fireplace and installed according to the fireplace manufacturer's instructions. [Ref. R1006.1 and R1006.4]

Intakes for exterior air may be located on the exterior of the dwelling or in naturally ventilated attics or crawl spaces. The IRC does not permit the installation of combustion air intakes in garages, basements, or mechanically ventilated spaces. Where combustion air openings are located inside the firebox, the exterior termination of the air intake cannot be higher than the firebox. Such an installation could result in combustion products being drawn to the outside through the air intake duct, creating a fire hazard. Exterior air intakes must be covered with a corrosion-resistant screen of ¼-inch mesh. [Ref. R1006.2]

MASONRY CHIMNEYS AND FIREPLACES

The prescriptive provisions of the IRC address structural support, approved materials, dimensions, and fire safety for masonry fireplaces. The code also prescribes construction details of masonry chimneys for proper drafting, weather protection, and safety from fire.

Design for masonry chimneys and fireplaces also must comply with other structural provisions of the code, including foundation requirements.

Footings

For masonry chimneys and fireplaces, the IRC requires concrete or solid masonry footings that are at least 12 inches thick and extend at least 6 inches beyond the face of the fireplace or foundation wall on all sides (Figure 11-2). **[Ref. R1001.2 and 1003.2]**

Seismic requirements

Masonry without reinforcing does not perform well in resisting the effects of earthquakes. Chimneys in particular, because of their height and narrow width, often suffer severe damage or collapse in significant seismic events. Proper anchorage to the dwelling structure and the

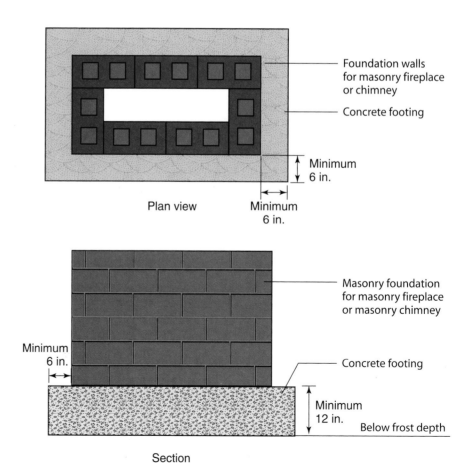

FIGURE 11-2 Footings for masonry chimneys and fireplaces

installation of reinforcing steel improve performance during earthquakes. The prescriptive seismic anchorage and reinforcing requirements in the IRC apply only to chimneys located in seismic design categories (SDCs) D_0, D_1, and D_2. There are no specific seismic requirements for chimneys in SDC "A," "B," or "C." Buildings in SDCs "E" and "F" require an engineered design in accordance with the *International Building Code* (IBC).

For masonry and concrete chimneys up to 40 inches wide with a single flue and located in SDC "D_0," "D_1," or "D_2," the IRC requires four continuous vertical No. 4 reinforcing bars. Two additional vertical bars are required for each additional flue or each additional 40 inches of width or fraction thereof. Vertical reinforcing must be enclosed in ¼-inch horizontal ties spaced not more than every 18 inches vertically. Two horizontal ties are required wherever there are bends in the vertical bars. In addition, masonry or concrete chimneys require lateral support through prescribed anchorage at every floor, ceiling, or roof more than 6 feet above grade (Figure 11-3). **[Ref. R1003.3, R1003.4, and Table R1001.1]**

Masonry fireplace details

Solid masonry or concrete firebox walls with minimum 2-inch-thick fire brick lining require a total thickness of not less than 8 inches, including the lining. Prescribed dimensions for the firebox, throat, and smoke

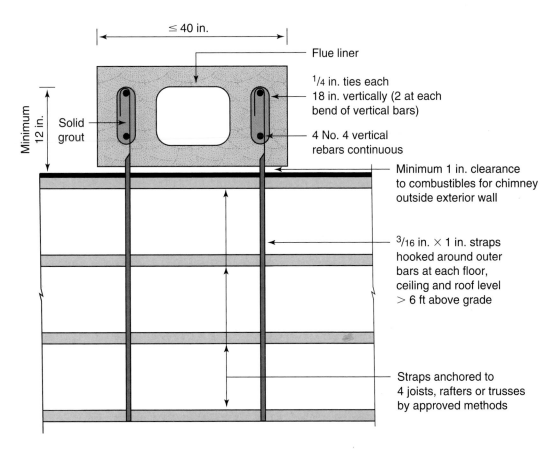

FIGURE 11-3 Masonry chimney seismic reinforcing and anchorage in SDCs "D_0," "D_1,"and "D_2"

chamber facilitate the proper discharge of smoke and products of combustion through the chimney and intend to prevent downdrafts. Steel or other noncombustible lintels supporting masonry above the firebox require minimum 4-inch bearing support at each end. The code requires an operable steel or cast iron damper located not less than 8 inches above the firebox opening that can be closed when the fireplace is not in use (Figure 11-4). **[Ref. R1001.5 to R1001.8]**

Hearth and hearth extension

The hearth is the floor of the firebox and is constructed to withstand the intense heat of a wood fire. The hearth extension projects from the front of the firebox to provide a noncombustible area of protection against radiant heat and any embers, sparks, or ash that may escape from the firebox. Hearths and hearth extensions must be constructed of concrete or masonry, supported by noncombustible materials, and reinforced to carry their own weight and all loads. The prescribed minimum thicknesses are 4 inches for the hearth and 2 inches for the hearth extension. When the firebox opening is raised at least 8 inches above the floor, the hazard of igniting underlying materials is reduced, and the code permits a hearth extension of ⅜-inch-thick brick, concrete, stone, tile, or other approved noncombustible material supported by combustible construction. For fireplace openings with areas less than 6 square feet, hearth extensions must project at least 16 inches out from the face of the fireplace and at least 8 inches beyond each side of the opening. For larger openings, the dimensions are a minimum 20 inches in front of the face and a minimum 12 inches beyond each side of the fireplace opening (Figure 11-5). **[Ref. R1001.9 and R1001.10]**

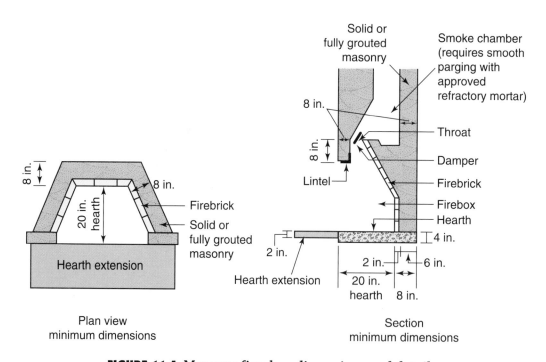

FIGURE 11-4 Masonry fireplace dimensions and details

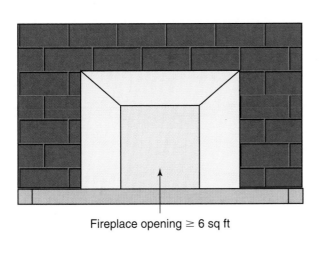

Fireplace opening ≥ 6 sq ft

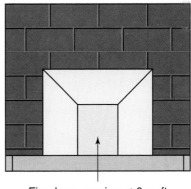

Fireplace opening < 6 sq ft

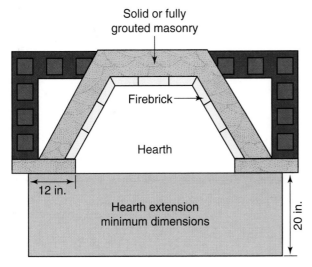

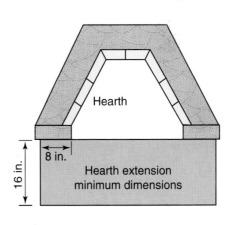

FIGURE 11-5 Hearth and hearth extensions for masonry fireplaces

Clearance to combustibles and fireblocking

To prevent fires caused by conductive and radiant heat transfer, the IRC prescribes clearances to wood and other combustible materials in proximity to masonry fireplaces and chimneys. In general, wood floor, wall, ceiling, and roof framing requires a minimum 2-inch air clearance from the masonry, though this dimension is increased to 4 inches on the back of the fireplace. Combustible sheathing, siding, flooring, trim, and gypsum board may abut the masonry fireplace and chimney sidewalls, provided that such combustible material maintains a distance of not less than 12 inches from the inside surface of the firebox and the inside surface of the flue (Figure 11-6). **[Ref. R1001.11 and R1003.18]**

Combustible mantels and trim placed directly on the front of the fireplace are not restricted when located more than 12 inches from the fireplace opening. Such materials located more than 6 inches but not greater than 12 inches from the fireplace opening are permitted but must not project more than ⅛ inch from the face of the masonry for each inch of separation from the edge of the fireplace opening. The code does not permit combustible trim within 6 inches of the opening (Figure 11-7). **[Ref. R1001.11]**

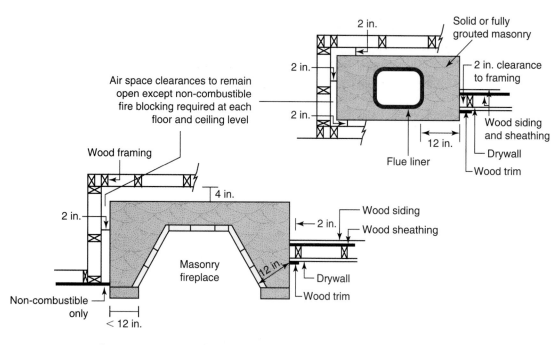

FIGURE 11-6 Clearances to combustible material from masonry fireplaces and chimneys

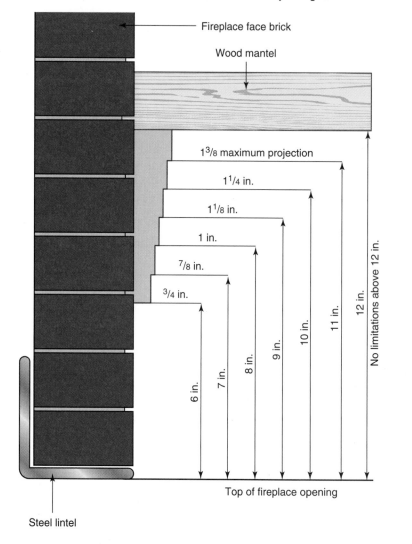

FIGURE 11-7 Clearances to combustible mantels and trim on the face of masonry fireplaces

The IRC requires fireblocking at prescribed intervals to stop the spread of fire in concealed spaces. This rule applies to the air spaces created by clearance to combustible materials around fireplaces and chimneys. In the case of masonry chimneys and fireplaces, the fireblocking must be noncombustible and must be installed at each floor and ceiling line. [Ref. R1001.12 and R1003.19]

Chimney dimensions and lining

The IRC prescribes minimum chimney wall thickness, flue size, and flue lining material to preserve the structural integrity of the chimney and to prevent the passage of damaging heat, flame, smoke, or products of combustion to adjacent construction or into the dwelling. Masonry chimney walls must be solid or fully grouted with a minimum nominal thickness of 4 inches. For masonry chimneys serving masonry fireplaces, the IRC requires an approved clay flue lining meeting the applicable material and installation standards, or another approved and listed chimney lining system. The IRC provides two methods for determining flue size. The first requires a minimum net flue cross-sectional area based on a ratio to the fireplace opening size. The ratio varies depending on the shape of the flue lining. The second option calculates flue size based on the fireplace opening area and the height of the chimney (Figure 11-8). [Ref. R1003.10 to R1003.15]

Chimney termination

To provide proper drafting, masonry chimneys must terminate at least 3 feet above the roof and at least 2 feet higher than any portion of a building within 10 feet (Figure 11-9). Flashing to weatherproof the chimney penetration at the roof must comply with the IRC roof covering and flashing requirements (see Chapter 7 of this publication). Where asphalt or wood shingles join the sides of a chimney, step flashing is required. For chimneys 30 inches wide or larger, the IRC also requires a cricket to direct water shed from the roof above around the sides of the chimney (Table 11-1 and Figure 11-10). [Ref. R1003.9, R1003.20, R903.2, and R905.2.8]

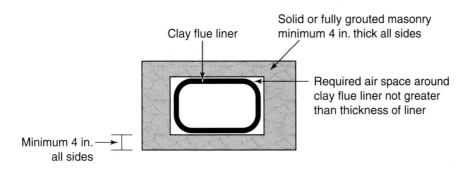

FIGURE 11-8 Masonry chimney wall thickness and flue lining

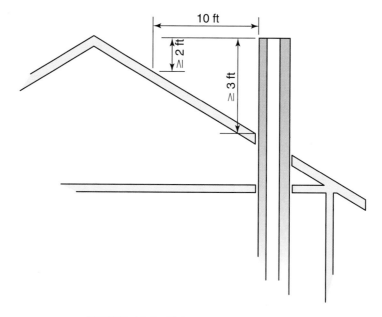

FIGURE 11-9 Chimney termination

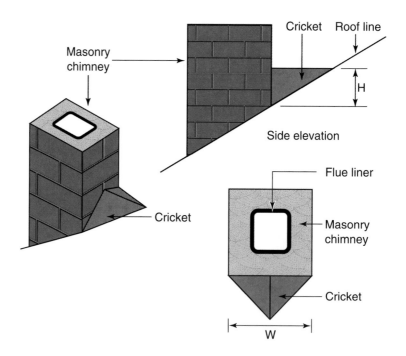

FIGURE 11-10 Cricket dimensions

TABLE 11-1 Cricket dimensions

Roof Slope	Height
12:12	½ of width
8:12	⅓ of width
6:12	¼ of width
4:12	⅙ of width
3:12	⅛ of width

MANUFACTURED CHIMNEYS AND FIREPLACES

Factory-built fireplaces and chimneys must be listed and labeled, tested in accordance with UL 127, and installed according to the conditions of the listing. The hearth extension dimensions prescribed for masonry fireplaces do not apply to manufactured fireplaces, which require hearth extensions to be installed in accordance with the listing of the fireplace. However, the IRC does require that hearth extensions be readily distinguishable from the surrounding floor area (Figure 11-11). **[Ref. R1004 and R1005]**

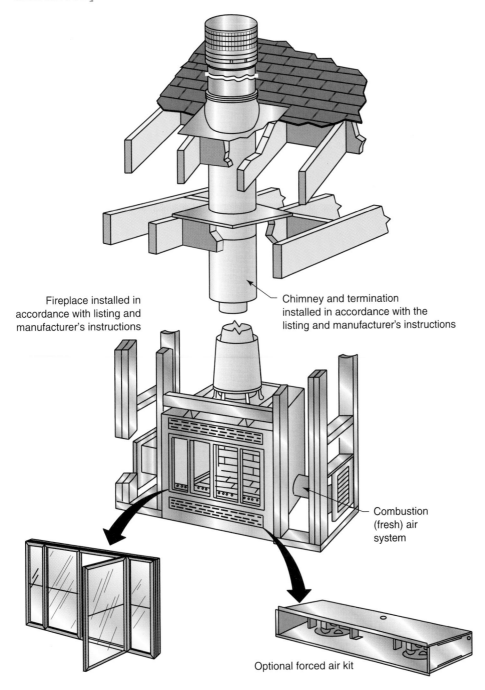

Fireplace installed in accordance with listing and manufacturer's instructions

Chimney and termination installed in accordance with the listing and manufacturer's instructions

Combustion (fresh) air system

Optional forced air kit

FIGURE 11-11 Factory-built fireplace and chimney

PART VI

Building Utilities

Chapter 12: Mechanical and Fuel-Gas

Chapter 13: Plumbing

Chapter 14: Electrical

CHAPTER 2

Mechanical and Fuel-Gas

The mechanical section of the *International Residential Code* (IRC) regulates the installation of permanently installed equipment and systems that control environmental conditions of a dwelling, such as comfort heating and cooling, including solid- or liquid-fuel appliances, duct systems, and ventilation systems. The fuel-gas section of the code regulates the installation of natural gas and liquid petroleum gas (LPG or LP gas) piping systems and the associated gas-fired appliances, including provisions for combustion air and venting of combustion products. Installations of mechanical systems that are not covered in the IRC must comply with the applicable provisions of the *International Mechanical Code* (IMC) or *International Fuel Gas Code* (IFGC). Discussion in this chapter will focus on common heating, ventilating, and air conditioning (HVAC) systems, gas-fired appliances, and gas piping systems.

APPLIANCES

Listing and labeling of appliances by qualified nationally recognized third party agencies, as mandated by the code, gives assurance that an appliance, when installed in accordance with the manufacturer's instructions, will function satisfactorily for the intended purpose and operate safely. The IRC requires the appliance to be installed and used in a manner consistent with the listing. For example, the listing may limit the use to a residential application in an indoor location. The required label is a factory applied nameplate identifying the manufacturer and the testing agency and providing other specified information. Labels for gas-fired appliances must indicate the hourly input rating in British thermal units per hour (Btu/h), the approved type of fuel (natural gas or LP gas), and the minimum clearances around the appliance (Figure 12-1). **[Ref. M1302, M1303 and G2404.3]**

Appliance installation and location

Appliance installation must conform to the requirements of the IRC and to the conditions of the appliance listing. **[Ref. R102.5, M1307.1, M1401.1, G2406.1, and G2408.1]**

You Should Know

Sizing of heating and cooling equipment and appliances must be in accordance with Air Conditioning Contractors of America (ACCA) manuals:

- Manual S, Residential Equipment Selection
- Manual J, Residential Load Calculation

Other calculation methodologies also may be approved. ●

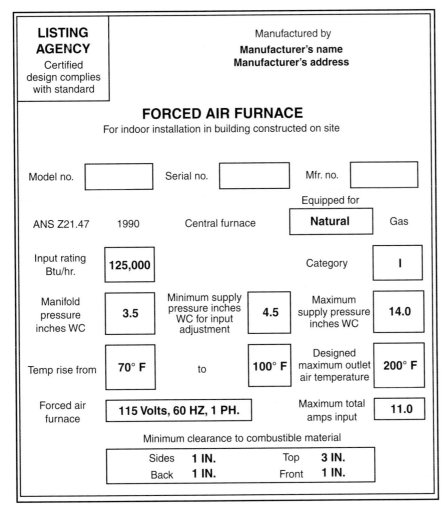

FIGURE 12-1 Appliance label

Clearances

The appliance listing and manufacturer's installation instructions generally determine minimum clearance to combustibles and minimum air clearance around the appliance for proper operation. In some instances, clearance to combustibles may be reduced with the application of noncombustible insulating materials, provided that such is not prohibited by the appliance listing. [Ref. M1306 and G2408.5]

Location limitations

In general, the IRC prohibits the installation of gas-fired appliances in sleeping rooms, bathrooms, toilet rooms, or storage closets, or in a space that opens only into such rooms or spaces. Such installation in small rooms with closed doors increases the risk of inadequate combustion air, improper operation, depleted oxygen levels, and exposure to carbon monoxide and other hazardous products of combustion. In addition, sleeping occupants are not alert to respond to developing hazardous conditions. Direct-vent appliances have sealed combustion chambers and draw all combustion air directly from the outside. Therefore, direct-vent appliances are permitted in these spaces and must be installed according to the manufacturer's instructions. Certain vented room heaters, fireplaces, and decorative appliances may be installed in bedrooms, bathrooms, or connecting spaces when the room contains the prescribed volume of combustion air.

With additional safeguards in place to isolate the appliance, the IRC does allow a gas-fired appliance, such as a furnace, boiler, or water heater, to be installed in a room or space that opens only into a bedroom or bathroom. Access to such space must be through a solid, weatherstripped, self-closing door, and all combustion air must be taken directly from the outdoors. The code also prohibits the space from being used for any other purpose, such as storage (Figure 12-2). [Ref. G2406.2]

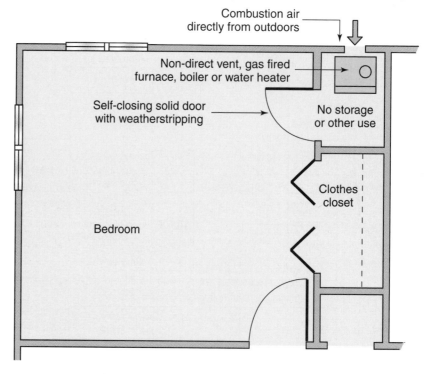

FIGURE 12-2 Gas appliance installed in a space that opens only into a bedroom

Appliances in garages

Vapors from gasoline and other flammable liquids often found in garages are heavier than air and may accumulate near the floor in sufficient concentration to ignite in the presence of a flame or spark. Accordingly, the IRC requires any ignition source of an appliance installed in a garage to be at least 18 inches above the floor unless the appliance is listed as flammable-vapor-ignition resistant. An *ignition source* is defined as a flame, spark, or surface capable of igniting flammable vapors or fumes, and includes appliance burners, burner igniters, and electrical switching devices. [Ref. M1307.3 and G2408.2]

Protection from impact

Accidental physical damage to an appliance or its fuel connection also creates a hazardous condition that may result in fire, explosion, or improper appliance operation, and the IRC requires protection from impact by vehicles. Although appliance installations in garages are common and are most likely to be affected by close proximity to cars and trucks, outdoor locations, particularly those adjacent to driveways, also may be vulnerable to vehicle impact and require protection. Protection may be achieved by installing bollards, curbs, or other approved barriers. Suspended appliances with sufficient clearance above the floor and appliances installed in an alcove out of the path of vehicle travel are not subject to impact and do not require additional barriers (Figure 12-3). The IRC also prohibits placing any stress on the connections of the fuel-gas piping system thereby reducing the possibility of damage to gas pipe fittings causing leaks or ignition of fuel gas. [Ref. M1307.3.1 and G2408.6]

Exterior installation

Appliances installed outdoors must be listed for exterior locations (or provided with protection from the weather) and be supported on a level concrete slab or other approved material extending at least 3 inches above grade. When suspended, such installations require a minimum 6-inch clearance above grade. The clearance-to-grade requirements also apply to appliances installed in crawl spaces. [Ref. M1305.1.4.1, and G2408.4]

Access to appliances

In addition to the appliance manufacturer's installation instructions, the IRC provides for adequate access and clearance to facilitate the service, repair, and replacement of appliances. All appliances require a minimum 30-inch by 30-inch working space in front of the controls. Access doors and passageways to appliances must be at least 24 inches wide and large enough to remove the largest appliance. Unless the listing of the appliance is more stringent, the code requires furnace compartments to be at least 12 inches wider than the appliance, with at least 3 inches clearance at the sides and back. [Ref. M1305]

Appliances in attics

As with other locations, appliances installed in attics require a sufficient access opening to remove the largest appliance. Minimum opening size is 30 inches by 20 inches. Because of the difficulty of accessing and

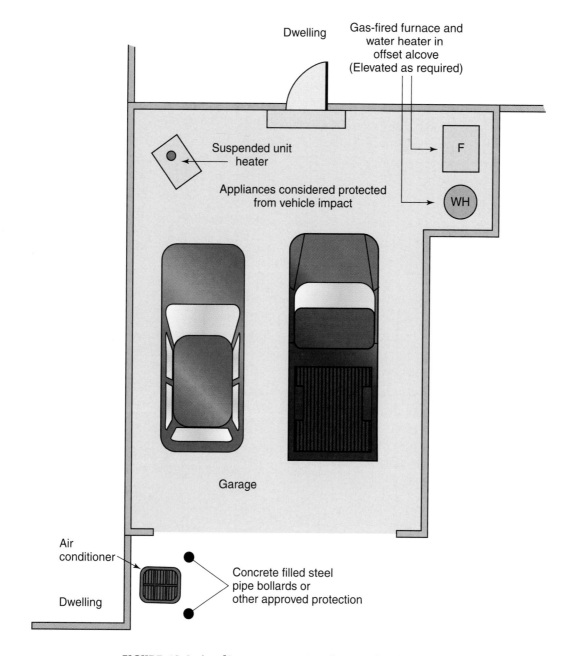

FIGURE 12-3 Appliance protection from vehicle impact

servicing appliances in attics unless they are located adjacent to the access opening, the code places restrictions on the distance to the appliance from the access opening and provides for a solid surface passageway not less than 24 inches wide. The length of the passageway is limited to not more than 20 feet unless there is a clear path at least 22 inches wide by 6 feet high, in which case the appliance may be located as much as 50 feet from the access opening. A solid platform is still required to satisfy the minimum 30-inch by 30-inch working space at the service side of the appliance. The IRC further requires a light fixture and receptacle outlet at the appliance location. The light must be controlled by a switch located at the access opening (Figure 12-4). Similar provisions apply to appliances installed under floors (crawl spaces). **[Ref. M1305.1.3]**

Condensate

The IRC regulates the proper disposal of condensate that results from either cooling or combustion processes. Cooling coils and evaporators installed in forced air furnaces as part of the air conditioning system generate condensate. High-efficiency category IV furnaces have low-temperature flue gases that also produce condensate in the vent. In many cases, both conditions exist in the same appliance. In general, condensate must drain to an approved location such as a floor drain. Where the appliance is installed in an upper story or an attic, where water leakage will cause damage to building components such as the drywall ceiling of the living space below, additional preventive measures are required.

The most common method to prevent water damage to construction materials because of a stoppage in the primary drain is to install an auxiliary drain pan below the appliance. In addition to prescribing the pan dimensions and materials, the IRC requires discharge to a conspicuous location to alert occupants of a problem. As an alternate to the aux-iliary drain pan, the code permits installation of a secondary ¾-inch drain line from the appliance's integral drain pan discharging to a conspicuous location. Acceptable alternatives to the auxiliary pan or secondary drain provide for automatic shut-down of the appliance when a stoppage in the drain occurs (Figure 12-5). **[Ref. M1411.3, M1411.4, and G2404.10]**

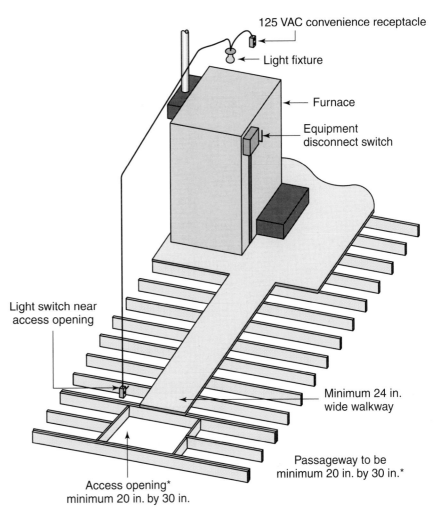

FIGURE 12-4 Attic installation requirements

- 125 VAC convenience receptacle
- Light fixture
- Furnace
- Equipment disconnect switch
- Light switch near access opening
- Minimum 24 in. wide walkway
- Passageway to be minimum 20 in. by 30 in.*
- Access opening* minimum 20 in. by 30 in.

*Large enough to allow removal of largest piece of equipment

EXHAUST SYSTEMS

Mechanical exhaust systems discharge airborne contaminants and moisture to the outside atmosphere. For other than whole-house ventilation systems, the IRC prohibits exhaust systems to terminate in an attic, soffit, ridge vent, or crawl space, to prevent damage to the structure from moisture. **[Ref. M1501]**

Clothes dryer exhaust systems

Exhaust systems for electric and gas clothes dryers must be installed in accordance with the appliance listing and the manufacturer's installation instructions. Dryer exhaust ducts convey moisture and, in the case of gas

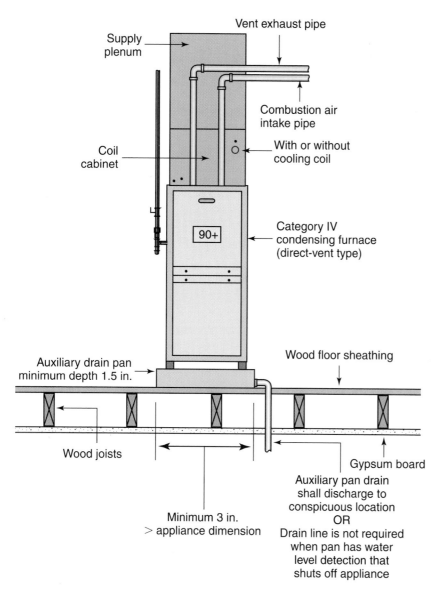

Supply plenum

Vent exhaust pipe

Combustion air intake pipe

Coil cabinet

With or without cooling coil

90+

Category IV condensing furnace (direct-vent type)

Auxiliary drain pan minimum depth 1.5 in.

Wood floor sheathing

Wood joists

Gypsum board

Auxiliary pan drain shall discharge to conspicuous location
OR
Drain line is not required when pan has water level detection that shuts off appliance

Minimum 3 in.
> appliance dimension

FIGURE 12-5 Auxiliary drain pan for condensate

dryers, combustion products to the outdoors. Because dryers discharge combustible lint, the code prescribes measures to reduce lint buildup, thereby reducing the hazard of a fire. In addition to conforming to the manufacturer's instructions, ducts must be 4 inches in diameter and constructed of smooth, rigid metal at least 0.016 inch thick. Joints must be assembled in the direction of air flow with no screws or connectors penetrating the duct. Maximum duct length is determined by the specified length requirements of the code, including reductions for fittings based on the degree and radius of the bend, or by the dryer manufacturer's instructions. The code limits connectors between the dryer and the rigid duct to a single piece of approved listed and labeled transition duct not more than 8 feet long. No connector can be concealed.

To prevent discharged air from reentering the building, dryer exhaust ducts must terminate outside at least 3 feet (or a distance specified by the manufacturer) from openings such as windows, doors, or ventilation intake locations. The IRC requires a backdraft damper and prohibits the installation of screens at the termination point (Figure 12-6). **[Ref. M1502 and G2439]**

Kitchen range hoods

Domestic open-top broiler units require an overhead metal exhaust hood unless the unit is equipped with a listed integral system. This hood is not required for other domestic kitchen cooking appliances. However, when installed, range hoods must comply with the manufacturer's instructions and the code requirements. For other than downdraft systems and listed ductless (recirculating) hoods, the IRC requires a single-wall, smooth, airtight duct of galvanized steel, stainless steel, or copper terminating outdoors with a backdraft damper. In addition, to prevent negative pressure and the accompanying adverse effects on mechanical appliances and systems within the dwelling, makeup air is required for high-velocity fans exceeding an exhaust capacity of 400 cfm (Table 12-1 and Figure 12-7). **[Ref. M1503 and M1505]**

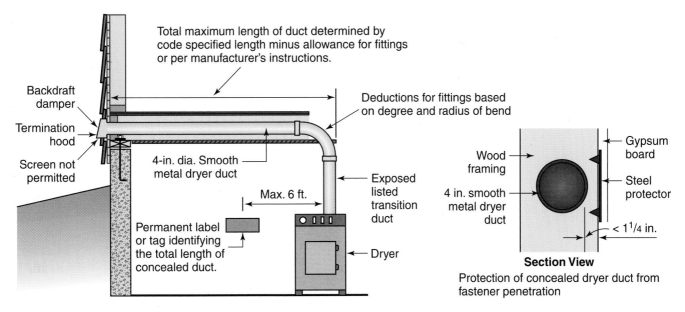

FIGURE 12-6 Clothes dryer exhaust system

TABLE 12-1 Minimum required exhaust rates

Area to Be Ventilated	Ventilation Rates
Kitchens	100 cfm intermittent or 25 cfm continuous
Bathrooms—toilet rooms	Mechanical exhaust capacity of 50 cfm intermittent or 20 cfm continuous

Bathrooms and toilet rooms with mechanical ventilation

Unless there is an openable window, bathrooms and toilet rooms require mechanical ventilation to exhaust air directly to the outdoors. The IRC prohibits the recirculating of the exhaust air within the dwelling. Ventilation rates must comply with Table 12-1. [Ref. M1507]

DUCT SYSTEMS

In order to effectively and safely circulate environmental air, duct systems must be fabricated of approved materials that meet the temperature, flame-spread, and smoke-developed ratings of the code. Factory-made ducts must be listed and labeled as complying with applicable standards of Underwriters Laboratories (UL). The IRC is concerned primarily with the burning characteristics of duct materials, including insulation, and the hazards related to the spread of smoke in the case of a fire. In addition to these fire-safety requirements, the IRC prescribes airtight joints and seams and installation to adequately ensure the structural integrity of the duct system (Figure 12-8). [Ref. M1601]

You Should Know

Gas-fired cooking appliances listed for use in commercial occupancies are not permitted within dwellings. •

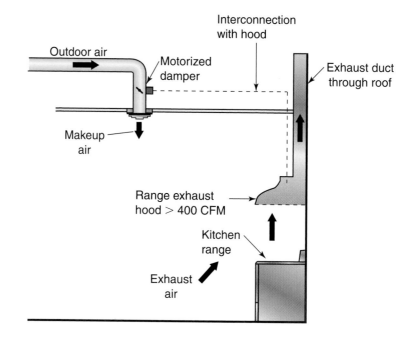

FIGURE 12-7 Required makeup air for kitchen exhaust hoods exceeding 400 cfm

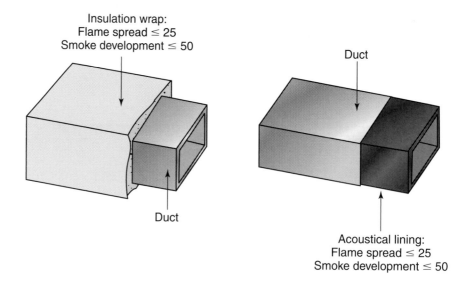

FIGURE 12-8 Duct insulation

The IRC permits isolated stud-wall cavities and panned solid-joist spaces to serve as return air ducts, provided that they are not part of a fire resistance rated assembly, do not convey return air from more than one floor level, and are properly fire blocked (Figure 12-9). **[Ref. M1601.1.1]**

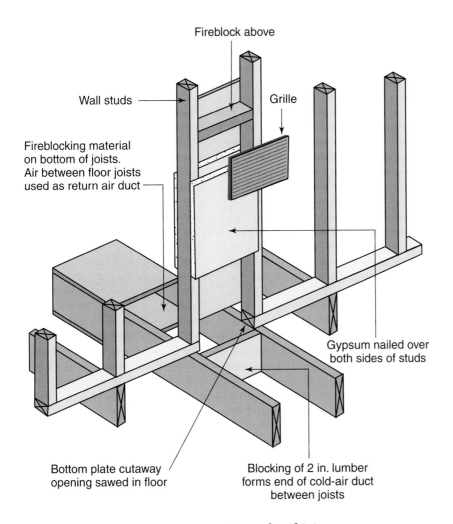

Fireblock above

Wall studs

Grille

Fireblocking material
on bottom of joists.
Air between floor joists
used as return air duct

Gypsum nailed over
both sides of studs

Bottom plate cutaway
opening sawed in floor

Blocking of 2 in. lumber
forms end of cold-air duct
between joists

FIGURE 12-9 Return air in stud and joist spaces

RETURN AIR

Return air is typically the air removed from an approved conditioned space and recirculated through the HVAC system. The routine operation of a dwelling may cause available return air to be lost through exhaust systems, appliance vents, and fireplace chimneys. Recognizing this, the IRC permits return air to be diluted or supplemented with fresh outdoor air through outdoor air inlets that must be screened.

The IRC restricts the sources of return air to prevent the circulation of unpleasant or noxious odors or other contaminants and to prevent negative pressures in confined spaces or appliance locations. [Ref. M1602 and G2442.5]

COMBUSTION AIR

Fuel-burning appliances require a supply of air for fuel combustion, draft hood dilution, and ventilation of the space in which the appliance is installed. Combustion air is important not only for proper operation of the appliance but for protecting occupants against the hazards of oxygen depletion and buildup of harmful combustion gases. The IRC references the combustion air requirements of the manufacturer for solid fuel-burning appliances and NFPA 31 for oil-fired appliances. Discussion here focuses on the specific combustion air requirements for gas-fired appliances. [Ref. M1701 and G2407.1]

Calculating the net free area of vents or grilles

When using louvers or grilles to satisfy combustion air requirements, the amount of net free area is as specified by the manufacturer, if known. If not known, the amount can be calculated as 75 percent of the gross area for metal louvers and 25 percent of the gross area for wood louvers (Figure 12-10). [Ref. G2407.10]

Combustion air from inside the building

Combustion air for gas-fired appliances may be obtained from an indoor space having a volume of at least 50 cubic feet per 1000 Btu/h input rating of all appliances being served within the space. The IRC also allows drawing combustion air from adjacent rooms through two permanent openings. One opening must be within 12 inches of the ceiling and one must be within 12 inches of the floor. The code requires each opening to have a free area of not less than 100 square inches and at least 1 square inch per 1000 Btu/h input rating of all appliances installed within the space (Figure 12-11). [Ref. G2407.5.2 and G2407.5.3]

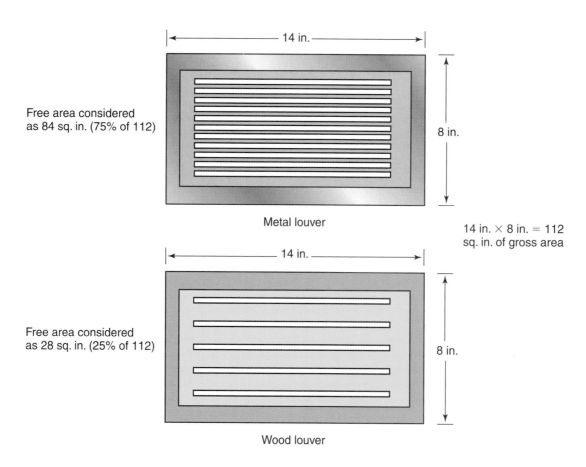

Metal louver

14 in. × 8 in. = 112 sq. in. of gross area

Wood louver

FIGURE 12-10 Louver net free area

Combustion air from outdoors

For gas-fired appliances, the IRC prescribes two methods for obtaining combustion air from the outdoors or from a space freely communicating with the outdoors, such as a ventilated attic. [**Ref. G2407.6**]

Outdoor combustion air obtained through two openings or ducts

In the first method for obtaining outdoor combustion air, two openings or ducts are required, one within 12 inches of the ceiling and one within 12 inches of the floor. Each vertical duct or direct opening to the outdoors requires a free area of at least 1 square inch per 4000 Btu/h of total input rating. Horizontal ducts require a larger cross-sectional free area of at least 1 square inch per 2000 Btu/h of total input rating.

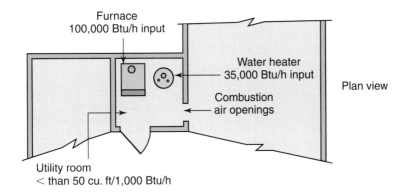

Furnace
100,000 Btu/h input

Water heater
35,000 Btu/h input

Plan view

Combustion
air openings

Utility room
< than 50 cu. ft/1,000 Btu/h

Example: Calculate indoor combustion air volume:

$$\frac{135,000}{1,000} \times 50 = \begin{array}{l} 6750 \text{ cu. ft} \\ \text{volume required} \end{array}$$

Example: Determine net free area for each combustion air opening
Total appliance input -135,000 Btu/h

$$\frac{135,000}{1,000} = \begin{array}{l} 135 \text{ sq in. net free} \\ \text{area per opening} \end{array}$$

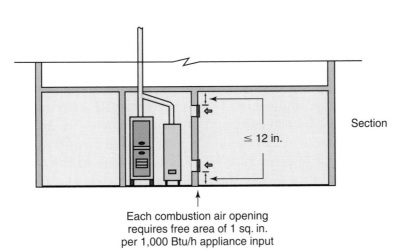

Section

≤ 12 in.

Each combustion air opening
requires free area of 1 sq. in.
per 1,000 Btu/h appliance input

FIGURE 12-11 Indoor combustion air from spaces on same story

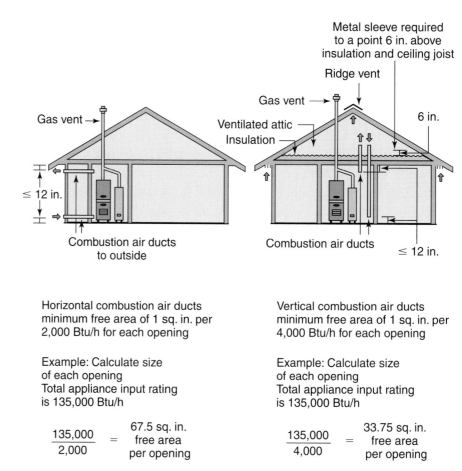

Horizontal combustion air ducts minimum free area of 1 sq. in. per 2,000 Btu/h for each opening

Vertical combustion air ducts minimum free area of 1 sq. in. per 4,000 Btu/h for each opening

Example: Calculate size of each opening Total appliance input rating is 135,000 Btu/h

Example: Calculate size of each opening Total appliance input rating is 135,000 Btu/h

$$\frac{135,000}{2,000} = \begin{array}{l} 67.5 \text{ sq. in.} \\ \text{free area} \\ \text{per opening} \end{array}$$

$$\frac{135,000}{4,000} = \begin{array}{l} 33.75 \text{ sq. in.} \\ \text{free area} \\ \text{per opening} \end{array}$$

FIGURE 12-12 Combustion air from outdoors through two openings

For combustion air ducts that terminate in an attic, the IRC requires the termination point to be not less than 6 inches above the top of the ceiling joists and insulation, and does not permit screens on the termination inlet (Figure 12-12). **[Ref. G2407.6.1]**

Outdoor combustion air obtained through single opening or duct

The IRC also permits combustion air for gas-fired appliances to be obtained through a single opening located within 12 inches of the ceiling when the size is increased to meet three criteria. The free area of the opening must be at least 1 square inch per 3000 Btu/h of the total appliance input rating and must be at least the sum of the areas of all vent connectors in the space. The code also prescribes minimum clearances around the appliances for free circulation of air (Figure 12-13). **[Ref. G2407.6.2]**

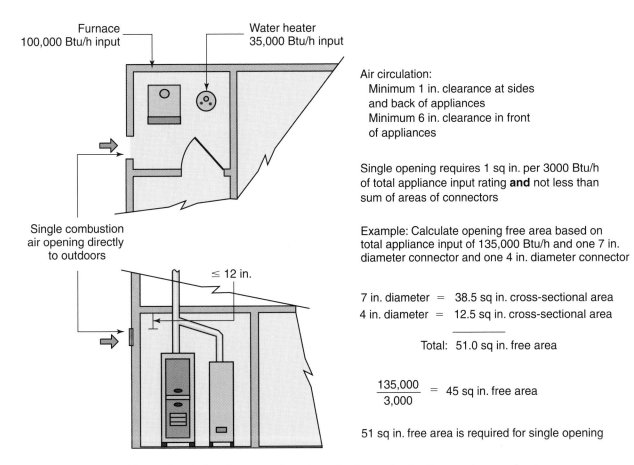

Furnace
100,000 Btu/h input

Water heater
35,000 Btu/h input

Air circulation:
   Minimum 1 in. clearance at sides
   and back of appliances
   Minimum 6 in. clearance in front
   of appliances

Single opening requires 1 sq in. per 3000 Btu/h
of total appliance input rating **and** not less than
sum of areas of connectors

Single combustion
air opening directly
to outdoors

Example: Calculate opening free area based on
total appliance input of 135,000 Btu/h and one 7 in.
diameter connector and one 4 in. diameter connector

≤ 12 in.

7 in. diameter = 38.5 sq in. cross-sectional area
4 in. diameter = 12.5 sq in. cross-sectional area
                            —————————
          Total: 51.0 sq in. free area

$$\frac{135,000}{3,000} = 45 \text{ sq in. free area}$$

51 sq in. free area is required for single opening

FIGURE 12-13 Outdoor combustion air through single opening or duct

VENTS

For gas-fired appliances, the code prescribes the methods for venting combustion products to the outside atmosphere. Other than plastic vents for category IV appliances, vents must be listed and labeled for use with the type of appliance.

Vent installation

The code addresses primarily two concerns of vent installation: separation from combustible materials to prevent ignition and protection of the vent from physical damage. The vent listing and manufacturer's instructions determine minimum clearances to combustibles. For vents passing through insulated attics, a 26-gauge sheet metal insulation shield that terminates at least 2 inches above the insulation is required. Where vents are installed in concealed locations through holes or notches in framing and are less than 1½ inches from the edge of the framing, the vents require protection from fastener penetration by installing a fastener shield plate of 16-gauge steel (Figure 12-14). **[Ref. G2426]**

Gas vent roof termination

The required termination height for gas vents not more than 12 inches in size and located at least 8 feet from a vertical wall is based on the roof pitch (Table 12-2 and Figure 12-15). Gas vents larger than 12 inches or less than

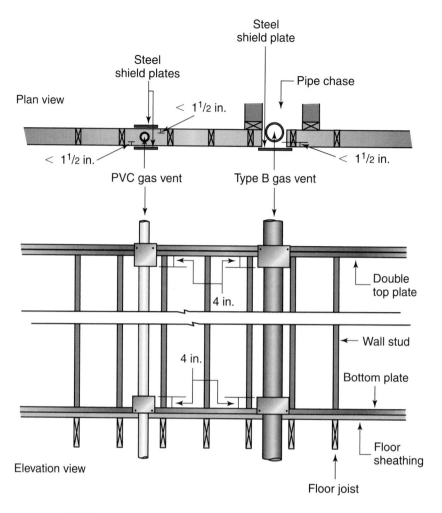

FIGURE 12-14 Vent protection from physical damage

TABLE 12-2 Gas vent termination height for listed caps 12 inches and smaller located at least 8 feet from a vertical wall

Roof Slope	Minimum Height (ft.) from Roof to Lowest Discharge Opening
Flat to 6/12	1.0
Over 6/12 to 7/12	1.25
Over 7/12 to 8/12	1.5
Over 8/12 to 9/12	2.0
Over 9/12 to 10/12	2.5
Over 10/12 to 11/12	3.25
Over 11/12 to 12/12	4.0
Over 12/12 to 14/12	5.0
Over 14/12 to 16/12	6.0
Over 16/12 to 18/12	7.0
Over 18/12 to 20/12	7.5
Over 20/12 to 21/12	8.0

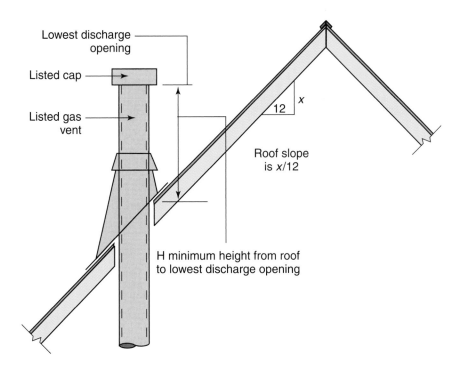

Lowest discharge opening

Listed cap

Listed gas vent

Roof slope is *x*/12

H minimum height from roof to lowest discharge opening

FIGURE 12-15 Gas vent roof termination

TABLE 12-3 Gas vent termination for direct-vent appliances

Appliance Btu/h Input Rating		Min. Clearance to Air Openings into Building (in.)	Min. Clearance Above Grade (in.)
Over	Not Over		
–	10,000	6	12
10,000	50,000	9	12
50,000	–	12	12

8 feet from a vertical wall must terminate at least 2 feet above the roof or any portion of a building within 10 feet horizontally. [Ref. G2427.6.3]

Direct-vent appliance vent termination

Direct-vent appliances draw all combustion air directly from outside. Direct-vent appliances often produce lower-temperature flue gases that may be vented through the exterior wall and terminate near the combustion air intake location. Clearances from vent terminations to building openings are less than would otherwise be required for non–direct vent appliances (Table 12-3 and Figure 12-16). [Ref. G2427.8]

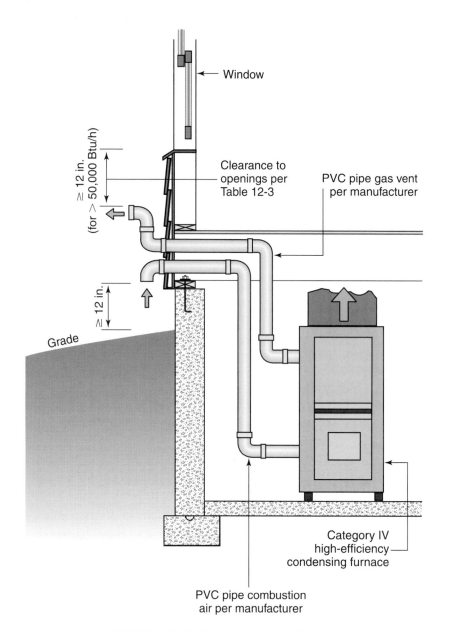

FIGURE 12-16 Direct-vent appliance

FUEL-GAS PIPING

The IRC provides for the design, materials, and safe installation of fuel-gas piping to serve gas-fired appliances.

Pipe sizing

Gas piping is sized to supply adequate volume to meet the demand of the connected appliances. Pipe size is determined based on a number of variables, including appliance input ratings, type and specific gravity of the gas, pipe material, length of pipe, inlet pressure, and pressure drop. The code includes pipe sizing tables and sizing equations but permits other approved methods for sizing gas pipe, including manufacturer's instructions and engineered methods. **[Ref. G2413]**

Piping materials

Approved gas piping materials include schedule 40 steel, approved seamless metallic tubing if gas used is not corrosive to the material, and corrugated stainless steel tubing (CSST). Approved plastic pipe, tubing, and fittings are permitted in exterior underground installations. Fittings and joint compounds when used must be compatible with the piping material and gas and approved for the specific use. [Ref. G2414]

Piping system prohibited locations

The IRC does not permit the installation of gas piping within an air duct, clothes chute, chimney, or gas vent or through any townhouse unit other than the unit being served. In addition, gas piping is not permitted to penetrate foundation walls below grade and must enter and exit a building at a point above grade. [Ref. G2415.1 and G2415.4]

Other installation requirements

Drilling and notching of wood floor, wall, and roof framing is limited to locations and dimensions as specified in Chapter 6 of this publication. Concealed piping installed through holes or notches in studs, joists, rafters, or similar members and which is less than 1½ inches from the nearest edge of the member must be protected by $^1/_{16}$-inch-thick nail shield plates. Schedule 40 black or galvanized steel gas piping resists penetration and does not require such protection. CSST gas tubing requires protection in accordance with the code and the manufacturer's installation instructions. [Ref. M1308 and G2415.5]

Above-ground piping outdoors requires a clearance of 3½ inches above ground and above roof surfaces. Protection from corrosion, such as painting or galvanizing, is required for exposed exterior ferrous metal piping. Underground piping, which is often polyethylene plastic, must be approved for the location. Galvanizing is not considered adequate protection from corrosion for underground steel pipe, which requires wrapping with approved material. Underground piping must be buried at least 12 inches deep. The IRC requires inspection and pressure testing of all fuel-gas piping systems before they are concealed or put into service. [Ref. G2415.7, G2415.9, and G2417]

Appliance connections

Rigid metallic pipe and fittings, CSST, and listed and labeled appliance connectors are approved for appliance connection to the gas piping system. Connectors are not allowed to pass through walls, floors, partitions, ceilings, or appliance housings (other than connectors to fireplace inserts with proper grommets in accordance with the manufacturer's instructions). For other than rigid metallic pipe, connector length is limited to no more than 6 feet (Figure 12-17). [Ref. G2422]

Shutoff valve

To facilitate service and replacement, each appliance requires an accessible shutoff valve located upstream of the connector in the same room and within 6 feet of the appliance. The code permits shutoff valves located as much as 50 feet from the appliance when installed at a manifold and clearly identified. (Figures 12-17 and 12-18). [Ref. G2420.5]

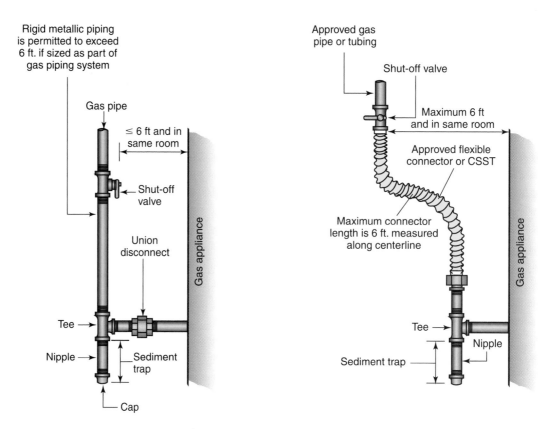

FIGURE 12-17 Gas appliance connection, shutoff valve, and sediment trap

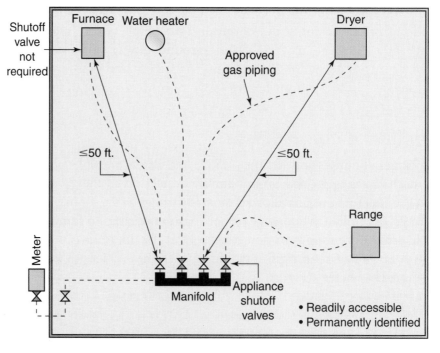

FIGURE 12-18 Appliance shutoff valves located at a manifold

TABLE 12-4 Support of fuel-gas piping

Steel Pipe, Nominal Size (in.)	Spacing of Supports (ft.)	Nominal Size of Tubing Smooth-Wall (in., outside diameter)	Spacing of Supports (ft.)
½	6	½	4
¾ or 1	8	⅝ or ¾	6
1¼ or larger (horizontal)	10	⅞ or 1 (horizontal)	8
1¼ or larger (vertical)	Every floor level	1 or larger (vertical)	Every floor level

Sediment trap

A sediment trap is generally required downstream of the shutoff valve and adjacent to the inlet of equipment. The IRC does not require sediment traps for illuminating appliances, ranges, clothes dryers, and outdoor grills (Figure 12-17). [Ref. G2419.4]

Piping support

Adequate piping support is necessary to prevent stresses on the pipe, fittings, and connections. See Table 12-4 for maximum spacing of supports for gas piping. CSST supports must follow the manufacturer's instructions. [Ref. G2424]

Plumbing

This chapter covers plumbing system design and installations typical of dwelling construction (Figure 13-1). Methods and materials outside the scope of the IRC must comply with the *International Plumbing Code* (IPC).

PIPING

While the IRC recognizes many types of approved piping materials, discussion here will focus on those materials most commonly encountered in construction of one- and two-family dwellings and townhouses. Materials must be third-party tested or certified as meeting the applicable IRC referenced standards. In order to perform as intended and to prevent damage or contamination, piping requires adequate support and protection from physical damage.

Protection from damage

Concealed piping installed through holes or notches in studs, joists, or rafters and less than 1½ inches from the nearest edge of the framing member requires protection from fastener penetration by shield plates. Protective shield

FIGURE 13-1 PEX water distribution tubing and PVC drain, waste, vent (DWV) piping

plates must be at least 16-gauge steel and cover the area where the pipe passes through the member. Shield plates must extend at least 2 inches above bottom plates and below top plates of wall framing. Cast iron and galvanized steel pipe are sufficiently resistant to penetration by nails or screws and do not require shield plate protection (Figure 13-2).

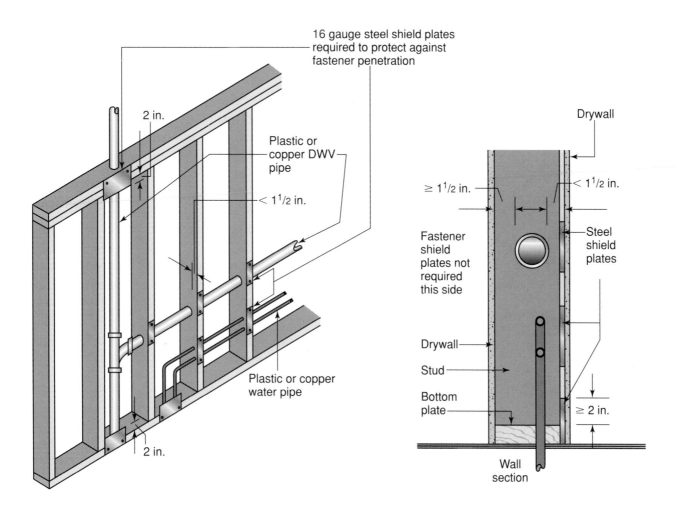

FIGURE 13-2 Physical protection of concealed piping

TABLE 13-1 Piping support

Piping Material	Max. Horizontal Spacing (ft.)	Max. Vertical Spacing (ft.)
ABS pipe	4	10
Cast-iron pipe, <10 ft. lengths	5	15
Cast-iron pipe, 10 ft. lengths	10	15
Copper or copper alloy pipe	12	10
Cross-linked polyethylene (PEX) pipe	2.67	10
PVC pipe	4	10

ABS = acrylonitrile butadiene styrene; PVC = polyvinyl chloride.

Limitations on boring and notching of structural members are covered in Chapter 6 of this publication. Pipes passing through footings or foundation walls require a pipe sleeve. **[Ref. P2603]**

Water, drain, and sewer piping must be protected from freezing. The IRC requires water service pipe to be buried at least 12 inches deep and at least 6 inches below the frost line. Minimum building sewer depth is determined by the building official and stipulated in the adopting ordinance. **[Ref. P2603.6]**

Piping support

Proper support of piping is important in maintaining alignment and slope, preventing sagging, and allowing for expansion and contraction. Piping installed underground requires continuous support on suitable bedding materials. Backfill over pipe must be free of debris, rocks, concrete, and frozen material. For above-ground installations, the IRC prescribes the maximum support spacing for horizontal and vertical piping based on the pipe material (Table 13-1). **[Ref. P2604 and P2605, Table 2605.1]**

Testing of piping systems

The IRC requires pressure testing of sewer, drain-waste-vent (DWV), and water piping systems to ensure that there are no leaks. The building sewer is water tested with not less than a 10-foot head of water. The DWV system requires a water test, also with 10-foot head of water, or an air test maintaining 5 pounds per square inch (psi) of pressure. The water supply system must be tested with a water pressure not less than the working pressure of the system. As an alternative, copper water piping may be tested with air at a minimum pressure of 50 psi. All tests must be maintained for 15 minutes without leakage. **[Ref. P2503]**

WATER SYSTEM

An approved potable-water supply is required for each dwelling unit, and the IRC prescribes methods for protecting and maintaining the system to deliver water that is safe for consumption by the occupants. The code also regulates design for adequate pressure and volume, suitable materials and fittings, and the location of shutoff valves.

Code Basics

DWV water test

- Piping system is filled with water to detect leaks.
- May test in stages or the entire system at one time.
- Requires 10 feet of water-filled vertical pipe above all piping being tested.
- 10-foot head pressure not required on top 10 feet of DWV piping (usually vent) in building.
- 10-foot column of water applies a pressure of 4.34 psi.
- In multistory buildings, lowest parts of the system tested may be under significantly higher pressure. •

Water service

Water service pipe is permitted in the same trench with a building sewer if the sewer pipe is listed for underground use within a building. For example, cast-iron or schedule 40 PVC DWV pipe is approved for this installation. For other types of building sewer pipe, such as polyethylene SDR-PR or PVC sewer and drain DR-PS pipe, the risk of contamination of the water supply is deemed greater, and the code requires separation. In these cases, the water service must have at least 5 feet of horizontal separation or be installed on a ledge at least 12 inches above and to one side of the highest point of the building sewer. [Ref. P2905.4.2]

Water supply system design criteria

The water system must be designed and installed to deliver adequate water volume and pressure for plumbing fixtures to operate efficiently and properly. The approved static pressure for the water service at the building entrance is 40–80 psi. Even if adequate pressure and volume are provided with a smaller pipe, the minimum size for water service pipe is ¾ inch. Water supply fixture unit values, developed length of piping, and water pressure determine pipe size for the distribution system. The IRC also limits flow rates and consumption for plumbing fixtures to conserve water. [Ref. P2903]

Valves

A main shutoff valve is required for every dwelling unit near the entrance of the water service. The valve must be of a full open type with a provision for drainage of the water distribution system. All valves must be accessible. Other required shutoff valve locations are shown in Table 13-2. [Ref. P2903.9]

TABLE 13-2 Shutoff valve locations

Appliance or Fixture	Valve Location	Valve Type
Water heater	Cold-water supply pipe at water heater	Full open
Lavatory	Each fixture supply pipe	Any approved type
Sink		
Water closet		
Bidet		
Bathtub	Not required	
Shower		
Hose bibb subject to freezing	Inside building*	Stop-and-waste type*

*Not required for frostproof hose bibb extending into heated building.

Dwelling unit fire sprinkler systems

The IRC plumbing provisions include a simple, prescriptive approach to the design of dwelling automatic fire sprinkler systems as an equivalent alternative to NFPA 13D systems. A dwelling fire sprinkler system requires less water when compared to NFPA 13 and 13R systems. The water supply requirements may be satisfied by a connection to a domestic water supply, a water well, an elevated storage tank, a pressure tank, or a stored water source with an automatically operated pump.

The code permits either a multipurpose sprinkler system, where domestic water is supplied to both sprinklers and plumbing fixtures, or a stand-alone sprinkler system. The code requires a rough-in inspection before piping is covered and a final inspection of the dwelling sprinkler system. Dwelling unit fire sprinkler systems are covered further in Chapter 9 of this publication. **[Ref. P2904]**

Water supply protection

The IRC requirements intend to protect the potable-water supply from contamination. Hose connections, boilers, heat exchangers, and lawn irrigation systems require listed backflow prevention devices suitable for the application. The simplest and most effective means of preventing contamination from drain water and the associated bacteria is through the use of an air gap. Sinks, lavatories, and bathtubs are examples of plumbing fixtures utilizing an air gap, which is the distance between the water outlet and the flood rim level of the fixture. The minimum air gap varies according to fixture type and application (Table 13-3 and Figure 13-3). A backflow preventer is not required between a stand-alone dwelling fire sprinkler system and the water distribution system. **[Ref. P2902, Table P2902.3.1, P2904.1]**

SANITARY DRAINAGE

Proper operation of the sanitary drainage system depends on adequate pipe size of approved materials, appropriate slope and support, transition fittings suitable for the location, adequate cleanouts, and proper venting.

TABLE 13-3 Minimum air gaps

Fixture	Minimum Air Gap	
	Away from a Wall (in.)	**Close to a Wall (in.)**
Effective openings greater than 1 inch	Two times the diameter of the effective opening	Three times the diameter of the effective opening
Lavatories and other fixtures with effective opening not greater than ½ inch in diameter	1	1.5
Over-rim bath fillers and other fixtures with effective openings not greater than 1 inch in diameter	2	3
Sinks, laundry trays, gooseneck back faucets, and other fixtures with effective openings not greater than ¾ inch in diameter	1.5	2.5

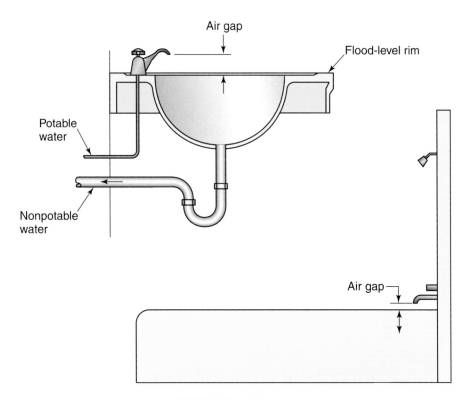

FIGURE 13-3 Air gaps

Connections and fittings

The code requires specific approved methods for joining and connecting drainage piping to provide a gas-tight system with optimal flow. Cast-iron pipe is typically connected with compression gasket joints or mechanical couplings. PVC DWV pipe joints are solvent cemented. These two different piping materials may be joined together with approved mechanical couplings. The code prohibits a reduction in size in the direction of flow. Fittings for change in direction must be suitable for the installation to maintain acceptable flow and reduce the possibility of stoppage. For example, sanitary tees are not permitted for vertical-to-horizontal or horizontal-to-horizontal transitions. Such changes in direction are commonly accomplished with wye, combination wye and eighth bend, or long sweep fittings (Figure 13-4). **[Ref. P3003 and P3004]**

Cleanouts

The IRC requires cleanouts at locations most susceptible to blockage and situated to accommodate drain cleaning equipment. Cleanouts are required in horizontal drain lines at each change of direction that is greater than 45 degrees. Where more than one change of direction occurs, only one cleanout is required in each 40 feet. A readily removable fixture, such as a water closet or a fixture trap of a sink, may serve as a cleanout. Accessible cleanouts are also required near the base of each vertical waste or soil stack and at the junction of the building drain and building sewer. Cleanouts must allow cleaning in the direction of the flow and provide working space clearance of at least 18 inches for pipes of 3 inches diameter or more, and 12 inches for smaller pipes (Figure 13-5). **[Ref. P3005.2]**

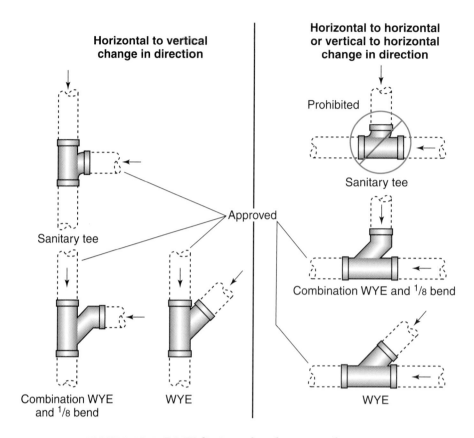

FIGURE 13-4 DWV fittings for change in direction

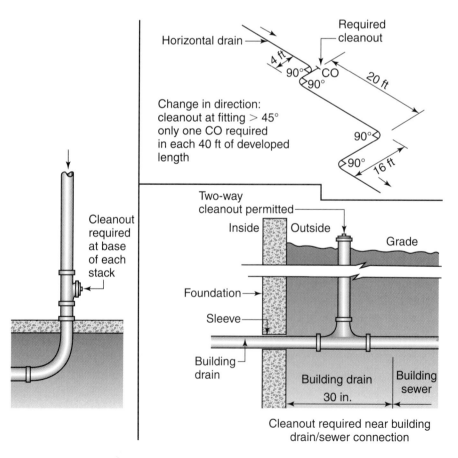

FIGURE 13-5 Cleanout locations

Minimum slope for horizontal drainage piping

The general rule for minimum pipe slope is ¼-inch vertical drop for each 12 inches of horizontal run (¼:12 or 2 percent). A minimum slope of ⅛ inch per foot (⅛:12 or 1 percent) is permitted for pipe sizes 3 inches or larger. [Ref. P3005.3]

Drain pipe sizing

In determining minimum pipe size, the IRC assigns a drainage fixture unit (d.f.u.) value to each plumbing fixture or group of fixtures. Each section of drainage pipe must be sized based on the total d.f.u of the fixtures draining through that section. The code provides separate tables; one for branches and stacks and one for the building drain and its branches. When compared with the tabular values for horizontal branches, values in the building drain table allow a greater d.f.u. capacity for any given size of pipe. Building drain capacity also increases as the slope increases up to a maximum slope of ½ inch per foot. Drain piping serving water closets must be at least 3 inches in diameter (Tables 13-4 through 13-6). [Ref. P3005.4, Table P3004.1, Table P3005.4.1, Table P3005.4.2]

> **EXAMPLE**
> Using the values in Tables 13-4 through 13-6, determine drainage fixture units and pipe size for each section of sanitary drainage pipe in Figure 13-6.

TABLE 13-4 Drainage fixture unit (d.f.u.) values for various plumbing fixtures

Type of Fixture or Group of Fixtures	d.f.u. Value
Bathtub, tub/shower, whirlpool or shower	2
Floor drain	0
Lavatory	1
Full-bath group with bathtub (with 1.6 gal. per flush water closet, and with or without shower head and/or whirlpool attachment on the bathtub or shower stall)	5
Half-bath group (1.6 gal. per flush water closet plus lavatory)	4
Kitchen group (dishwasher and sink with or without garbage grinder)	2
Laundry group (clothes washer standpipe and laundry tub)	3
Multiple-bath groups:	
1.5 baths	7
2 baths	8
2.5 baths	9

TABLE 13-5 Maximum fixture units allowed to be connected to branches and stacks

Nominal Pipe Size (in.)	Any Horizontal Fixture Branch	Any One Vertical Stack or Drain
1½	3	4
2	6	10
2½	12	20
3	20	48

Water closets are not permitted on drain lines less than 3 inches.

TABLE 13-6 Maximum fixture units allowed to be connected to building drain, building drain branches, or building sewer

Diameter of Pipe (in.)	Slope per ft.		
	in.	¼ in.	½ in.
2	–	21	27
3	36	42	50

Note: Water closets are not permitted on drain lines smaller than 3 inches.

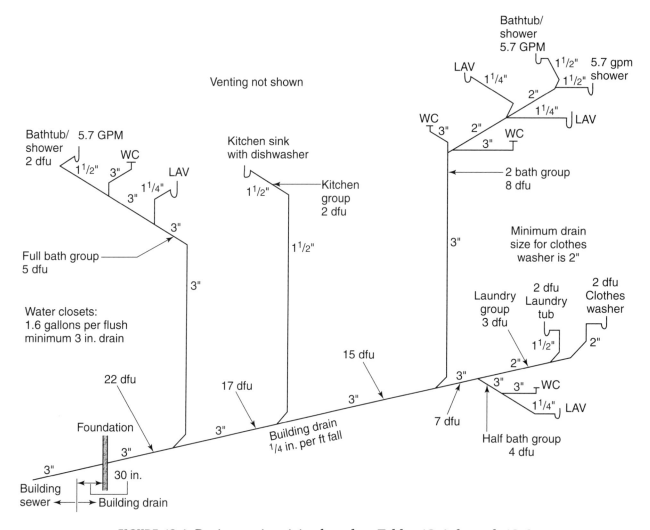

FIGURE 13-6 Drainage pipe sizing based on Tables 13-4 through 13-6

VENTING SYSTEMS

Venting provides air to equalize pressure for proper operation of the drainage system, prevents siphoning of trap seals, and allows the safe escape of sewer gases to the outside atmosphere. To protect the trap seal, the IRC requires venting for every trap and trapped fixture. At least one vent pipe to the outdoors is required and must be sized based upon the size of the building drain.

Vent connections and grades

Vents require adequate slope and alignment to allow moisture and condensation to drain back to the soil and waste pipes. To prevent a stoppage in a dry vent, which would likely go unnoticed, the dry vent connection to a horizontal drain must be above the centerline of the drain. If plumbing is roughed in for future fixtures, a vent must be installed to serve those fixtures. [Ref. P3104]

Fixture vents

For other than self-siphoning fixtures such as water closets, the IRC limits the distance from the trap to the vent and the slope of the fixture drain in order to keep the vent open above the flow line. The total fall of a fixture drain to the vent connection cannot exceed one pipe diameter, and the vent connection is not permitted to be below the trap weir (Table 13-7 and Figure 13-7). [Ref. P3105]

Wet venting

Wet venting is based on the principle that pipe serving as a drain for a fixture or fixtures will have adequate amounts of air to concurrently serve as the vent for downstream fixtures. The section of drainage pipe serving as a wet vent must be adequately sized and is limited to a lower d.f.u. load than would otherwise be allowed for drainage piping. Horizontal wet venting is permitted for fixtures of one or two bathroom groups located on the same floor (Table 13-8 and Figure 13-8). [Ref. P3108]

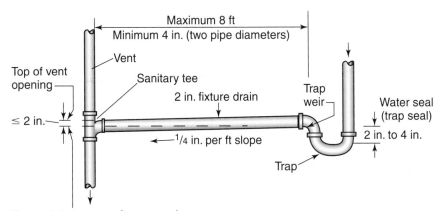

FIGURE 13-7 Fixture vent

TABLE 13-7 Maximum distance of fixture trap from vent

Size of Trap (in.)	Slope (in. per ft.)	Distance from Trap (ft.)
1¼	¼	5
1½	¼	6
2	¼	8

TABLE 13-8 Wet vent size

Wet Vent Pipe Size (in.)	Drainage Fixture Unit (d.f.u.) Load
1½	1
2	4
3	12

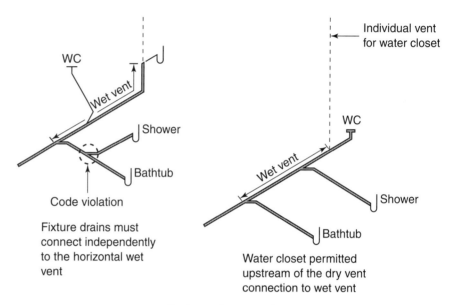

Horizontal wet venting

FIGURE 13-8 Wet venting

Island fixture venting

Island venting is permitted for lavatories and kitchen sinks, including connections for dishwashers and food waste disposers (Figure 13-9). [Ref. P3112]

Vent pipe sizing

To provide adequate volume of air to the drainage system, the IRC requires the diameter of vent piping to be at least half of the required diameter of the drain served, but never less than 1¼ inches. Vents greater than 40 feet long require an increase of one pipe size.

Vent termination

For sloped roofs not used for other purposes, vent pipes must terminate not less than 6 inches above the roof and 6 inches above the anticipated snow accumulation (Figure 13-10). To prevent noxious gas or odors from entering the building, the IRC prescribes minimum clearances between vent terminations and doors, windows, or air intake openings (see Chapter 10, Figure 10-3). [Ref. P3103]

You Should Know

A plumbing vent must extend at least 7 feet above a roof that is used for purposes other than weather protection, such as a flat roof used as a deck or other walking surface. ●

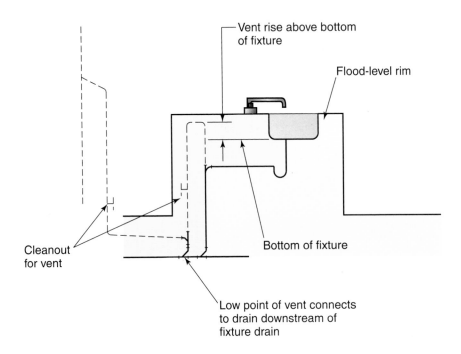

Island fixture vent

FIGURE 13-9 Island venting

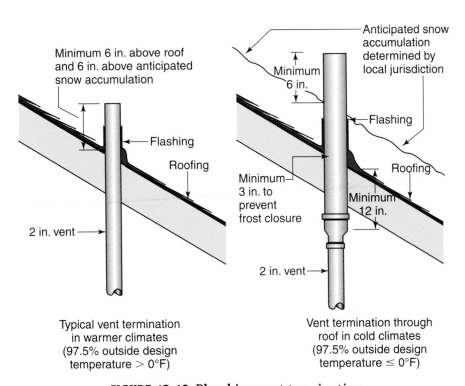

FIGURE 13-10 Plumbing vent termination

Air admittance valves

An air admittance valve is a one-way valve designed to allow air into the plumbing drainage system as water drains from the fixture, creating a negative pressure in the piping. The device pulls air from the room in which the fixture is located and, except for the one required

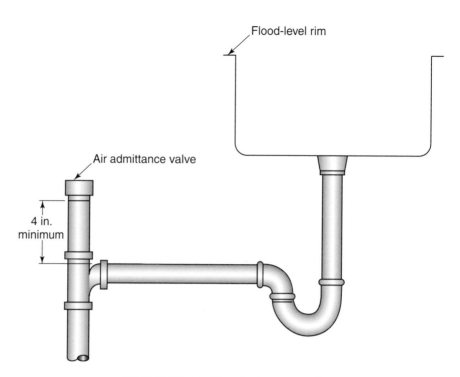

FIGURE 13-11 Air admittance valve

through-the-roof vent, is an alternative to venting to the outside air. The valve is designed to close by gravity and seal the terminal under no-flow conditions to prevent introduction of sewer gas into the dwelling. The space containing the air admittance valve requires access and ventilation. Use of air admittance valves may be an alternative to island venting of kitchen sinks or lavatories, or for venting other isolated fixtures (Figure 13-11). **[Ref. P3114]**

PLUMBING FIXTURES

Fixtures, faucets, and fixture fittings must have smooth, impervious surfaces and comply with the applicable referenced standards. The IRC includes requirements for receptors, strainers, valves, and features to promote usable and sanitary fixtures and prevent contamination of the potable-water supply.

Laundry standpipes

Standpipe design accommodates the high rate of discharge from clothes washers. Standpipes must extend at least 18 inches and not more than 42 inches above the trap weir (Figure 13-12). **[Ref. P2706.2]**

Dishwashers

The IRC provisions allow a kitchen sink, food grinder, and dishwasher to discharge through a single 1½-inch trap. The dishwasher drain may discharge through the food grinder or it may connect to the sink tailpiece with a wye fitting. The dishwasher waste line must be increased to ¾ inch

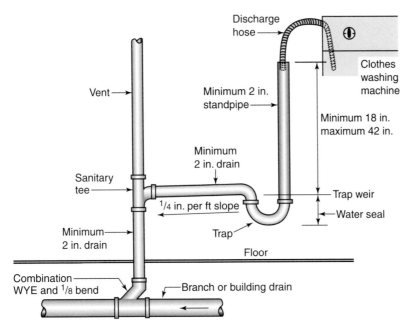

FIGURE 13-12 Laundry standpipe

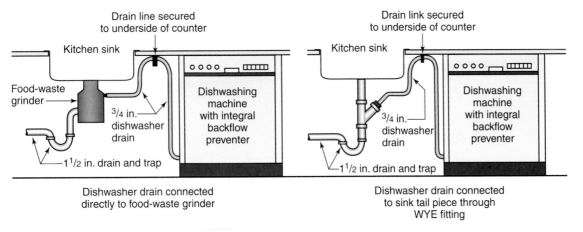

FIGURE 13-13 Dishwasher drain

diameter and be fastened securely to the underside of the counter to minimize the potential for wastewater backflow into the dishwasher (Figure 13-13). [Ref. P2717]

Protection against scalding

The IRC requires approved temperature control devices on the water outlets of bathing fixtures and bidets to prevent scalding (Table 13-9). [Ref. P2708.3, P2713.3, and P2721.2]

Showers

Minimum shower compartment dimensions are 30 inches by 30 inches. If the shower area is at least 1300 square inches, the minimum width may be reduced to 25 inches. Hinged shower doors must open outward, with a finished access width of at least 22 inches (see Chapter 10, Figure 10-6). [Ref. P2708]

TABLE 13-9 Temperature control at fixture water supply outlets

Fixture	Max. Temperature	Approved Device	Standard
Shower or tub/shower combination	120°F	Pressure-balance control valve	ASSE 1016 or CSA B125
		Thermostatic-mixing control valve / Combination pressure-balance/thermostatic-mixing control valve	
Bathtub or whirlpool bathtub	120°F	Water-temperature-limiting device	ASSE 1070
Bidet	110°F		

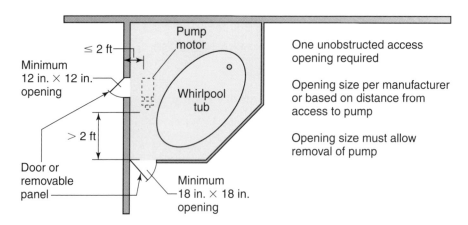

FIGURE 13-14 Access to whirlpool pump

Whirlpool bathtubs

Adequate access is necessary to service or replace the whirlpool circulation pump. The means and dimensions for access must follow the manufacturer's installation instructions and the code requirements. In all cases, the access opening must be unobstructed and large enough to remove the pump (Figure 13-14). **[Ref. P2720]**

FIXTURE TRAPS

Traps provide a water seal with a depth of 2 inches to 4 inches to prevent sewer gases from entering the building. Floor drains require a trap-primer or deep-seal design to prevent the loss of their water seal by evaporation (Figure 13-15). **[Ref. P3201.2]**

WATER HEATERS

The IRC prescribes location, installation, and safety requirements for water heaters in the plumbing, fuel-gas, and electrical chapters of the code. The plumbing section covers connections to the water supply, drain pan installations, and temperature relief valve provisions. As in the

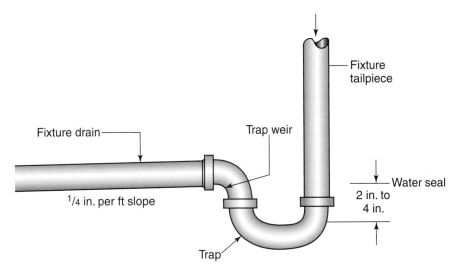

FIGURE 13-15 Fixture trap details

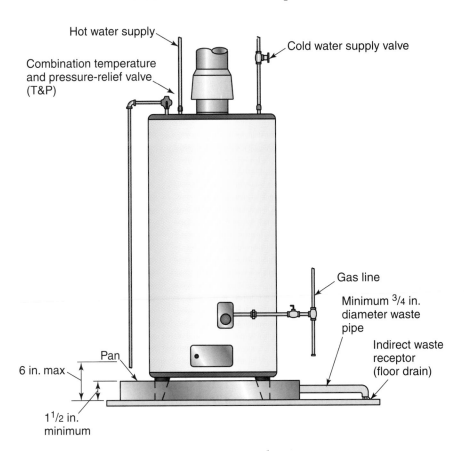

FIGURE 13-16 Water heater pan

fuel-gas provisions, water heaters installed in garages must have ignition sources elevated at least 18 inches above the garage floor. For buildings located in seismic design categories (SDCs) D_0, D_1, and D_2, and townhouses located in SDC "C," water heaters require anchorage to the walls of the structure to resist horizontal earthquake forces. Where leaking water will cause damage, the IRC requires that the water heater be installed in a galvanized steel or listed pan that drains to an approved location, typically a floor drain or other indirect receptor (Figure 13-16). [Ref. P2801 and P2803]

Electrical

The electrical chapters of the *International Residential Code* (IRC) are based on the *2008 National Electrical Code* (NEC) (NFPA 70-2008), published by the National Fire Protection Association (NFPA). The IRC provisions cover electrical installations normally occurring in the construction of one- and two-family dwellings and townhouses. This does not prevent the use of other wiring methods and materials that conform to the NEC requirements. Discussion in this chapter will focus on commonly encountered electrical installations for services, branch circuits, devices, and fixtures.

ELECTRICAL SERVICES

The scope of the IRC covers services for 120/240-volt, single-phase systems not more than 400 amperes. The serving utility company delivers electrical power to the service entrance conductors, which in turn conduct the energy to the main service disconnect. The service distributes electricity to the premises wiring system. Only one service is permitted for one- and two-family dwellings.

Equipment location

The electrical service equipment is typically a cabinet mounted to an interior wall and containing a panelboard with a main disconnect that shuts off all power to the building, the circuit breakers serving the branch circuits and related equipment. The service disconnect must be readily accessible and, when installed inside, must be near to where the service conductors enter the building. Working space in front of equipment, including service panels or subpanels, must be at least 30 inches wide by 36 inches deep with headroom of at least 6 feet 6 inches, and have a light source nearby. The spaces above and below the panel are dedicated to the electrical installation. This means, for example, that pipes or ducts cannot be located above the panel. The IRC also does not permit the installation of electrical panels, service disconnects, or circuit breakers in clothes closets or bathrooms (Figure 14-1). [**Ref. E3405 and E3601.6**]

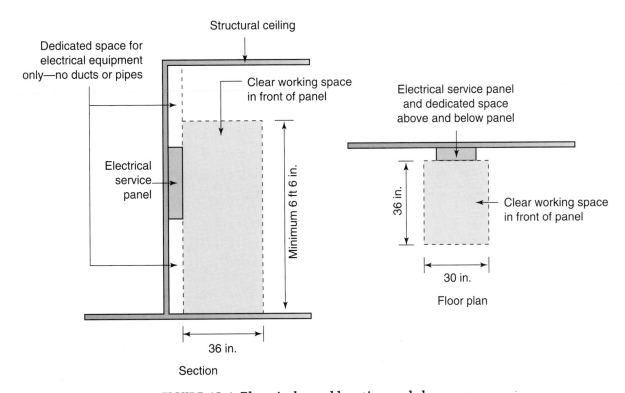

FIGURE 14-1 Electrical panel location and clearance

Electrical service size and rating

The minimum service rating for single-family dwellings is 100 amperes. The code permits a minimum service rating of 60 amperes for other installations. The ampacity of ungrounded service conductors and the rating of the service disconnect may not be less than the load served. The minimum service load for a single-family dwelling is calculated according to Table 14-1. **[Ref. E3602]**

Service conductor size

In the more precise language of the code, hot current-carrying wires are referred to as *ungrounded conductors*, and the neutral wires are referred to as *grounded conductors*. Table 14-2 identifies the minimum size of the ungrounded service conductor to provide the required ampacity based on the wire material and type of insulation. The grounded (neutral) conductor ampacity must be at least the maximum unbalance of the load and its size not less than the required minimum

TABLE 14-1 Minimum service load calculation for a single-family dwelling

Loads and Procedure				Volt-amperes
General lighting and general use receptacle outlets	3 volt-amperes	X	Floor area (sq. ft.)	_____ VA
				Plus
All 20-ampere-rated small appliance and laundry circuits	1500 volt-amperes	X	Number of circuits	_____ VA
				Plus
Appliances	The nameplate volt-ampere rating of all permanent or dedicated appliances:			
			Ranges	_____ VA
			Ovens	_____ VA
			Cooking units	_____ VA
			Clothes dryers	_____ VA
			Water heaters	_____ VA
			Subtotal	_____ VA
				Plus
Apply the following demand factors to the above subtotal:	100%	X	First 10,000 volt-amperes	_____ VA
				Plus
	40%	X	Portion in excess of 10,000 volt-amperes	_____ VA
				Plus
Air conditioning	Total of the nameplate rating(s) of the air-conditioning and cooling equipment			_____ VA
			Total volt-amperes	_____ **VA**
Total load in amperes	**Volt-ampere sum**	÷	**240 volts =**	_____ **AMPS**

TABLE 14-2 Service conductor and grounding electrode conductor size

| Conductor Types and Sizes* | | | Min. Grounding Electrode Conductor Size | |
Copper (AWG)	Aluminum and Copper-Clad Aluminum (AWG)	Allowable Ampacity, Max. Load (amps)	Copper (AWG)	Aluminum (AWG)
4	2	100	8	6
3	1	110	8	6
2	1/0	125	8	6
1	2/0	150	6	4
1/0	3/0	175	6	4
2/0	4/0 or two sets of 1/0	200	4	2
3/0	250 kcmil or two sets of 2/0	225	4	2
4/0 or two sets of 1/0	300 kcmil or two sets of 3/0	250	2	1/0
250 kcmil or two sets of 2/0	350 kcmil or two sets of 4/0	300	2	1/0
350 kcmil or two sets of 3/0	500 kcmil or two sets of 250 kcmil	350	2	1/0
400 kcmil or two sets of 4/0	600 kcmil or two sets of 300 kcmil	400	1/0	3/0

*THHN, THHW, THW, THWN, USE, RHH, RHW, XHHW, RHW-2, THW-2, THWN-2, XHHW-2, SE, USE-2.
Note: Service conductors in parallel sets of 1/0 and larger are permitted in either a single raceway or in separate raceways. Grounding electrode conductors of size 8 AWG require protection with conduit. Grounding electrode conductors of size 6 AWG require protection with conduit or must closely follow a structural surface for physical protection.

grounding electrode conductor size. Though in practice a rule of thumb has held that the neutral may be as much as two sizes smaller than the ungrounded service conductors, the code requires that the size be calculated. The grounding electrode system is discussed later in this chapter, but the size of the grounding electrode conductor is based on the size of the service entrance conductors, as shown in Table 14-2. **[Ref. E3603, Table E3603.1]**

GROUNDING

The grounding system provides a fault current path to earth. Grounding conductors from equipment, enclosures, and devices are bonded to the grounded (neutral) conductor and grounding electrode conductor at the service equipment. The electrical system is connected to the earth through the grounding electrode conductor(s) connected to one or more approved grounding electrodes in contact with the earth. **[Ref. E3607]**

Grounding electrode system

Commonly used grounding electrodes include underground metal water pipe, concrete-encased reinforcing bar, and approved ground rods. The electrodes, when present, are bonded together to form the grounding electrode system. An electrode of underground metal water pipe must have at least 10 feet in contact with the ground, with connection to the grounding electrode conductor within 5 feet of entrance into the building. Interior metal water piping more than 5 feet from the entrance into the building is not considered part of the grounding electrode system and cannot be used as a conductor to connect electrodes. Where underground metal water pipe is available and used as an electrode, the code requires at least one additional electrode.

Concrete-encased grounding electrodes (often called an *Ufer ground*) have proven very effective and do not require supplemental electrodes. The effectiveness of this method relies on the conductive characteristics of the concrete, the increased surface area in contact with the soil, and the ability of concrete to absorb and retain moisture. This type of electrode may be 20 feet of ½-inch reinforcing bar or 20 feet of 4 AWG bare copper wire encased in a concrete footing or foundation in contact with the ground. At least 2 inches of concrete cover is required. If such reinforcing is present in a concrete footing, the code requires it to be used as part of the grounding electrode system, though no more than one concrete encased electrode is required.

One or more rod, pipe, ring, or plate grounding electrodes, sometimes referred to as "made" electrodes, are required under two circumstances: where there are no other electrodes available or where underground water pipe is the only other grounding electrode present. This requirement is typically satisfied with a listed copper-clad ground rod ½ inch in diameter by 8 feet long driven into the soil. Unless testing after installation determines that a single ground rod has a resistance of 25 ohms or less, the code requires two such electrodes spaced not less than 6 feet apart.

Grounding electrode conductors are sized based on the size of the ungrounded service conductors, as shown in Table 14-2. The code generally does not permit splicing of grounding electrode conductors run to one or more grounding electrodes. Connections to electrodes require approved fittings to ensure an effective grounding path. Grounding electrode conductors require protection in conduit or by other approved means if exposed to potential physical damage (Figure 14-2). **[Ref. E3608, E3610, and E3611]**

Bonding

The code provides for the bonding of metal parts associated with the electrical system to provide an effective path for fault current. A main bonding jumper, such as the green machine screw or green insulated conductor supplied with the service enclosure, is required to connect the grounded (neutral) conductors to the service equipment enclosure and the equipment grounding conductors. As a general rule, the main service disconnect enclosure is the only location where the code permits connection of the grounding system to the grounded (neutral) conductors. They are isolated from each other elsewhere to prevent creating a parallel fault path.

Bonding of the metal water piping system to the service equipment enclosure with an appropriately sized bonding conductor is required in order to provide a fault path should the water pipe become energized

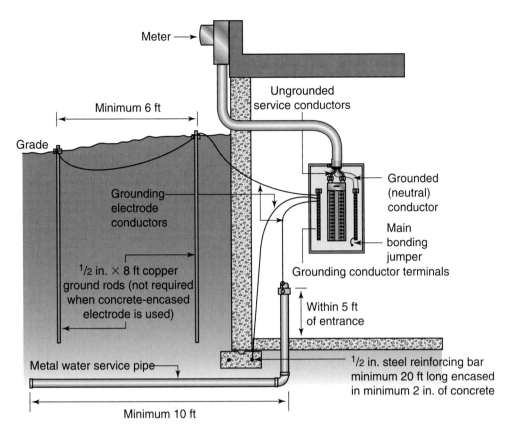

Meter →

Ungrounded
service conductors

Minimum 6 ft

Grade

Grounding
electrode
conductors

¹/₂ in. × 8 ft copper
ground rods (not required
when concrete-encased
electrode is used)

Grounded
(neutral)
conductor

Main
bonding
jumper

Grounding conductor terminals

Within 5 ft
of entrance

Metal water service pipe

¹/₂ in. steel reinforcing bar
minimum 20 ft long encased
in minimum 2 in. of concrete

Minimum 10 ft

FIGURE 14-2 Grounding electrode system

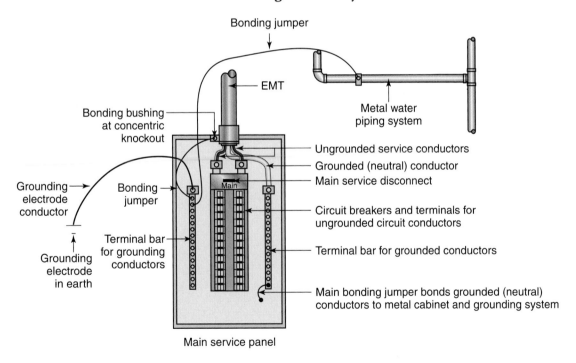

Bonding jumper

Bonding bushing
at concentric
knockout

EMT

Metal water
piping system

Grounding
electrode
conductor

Bonding
jumper

Ungrounded service conductors
Grounded (neutral) conductor
Main service disconnect

Grounding
electrode
in earth

Terminal bar
for grounding
conductors

Circuit breakers and terminals for
ungrounded circuit conductors

Terminal bar for grounded conductors

Main bonding jumper bonds grounded (neutral)
conductors to metal cabinet and grounding system

Main service panel

FIGURE 14-3 Bonding at the service panel

through accidental contact with hot wires. The code does not permit the
interior water piping system to be used as a ground, a grounding electrode, or a conductor for the grounding electrode system. The points of
attachment of the bonding jumper must be accessible (Figure 14-3).
[Ref. E3609]

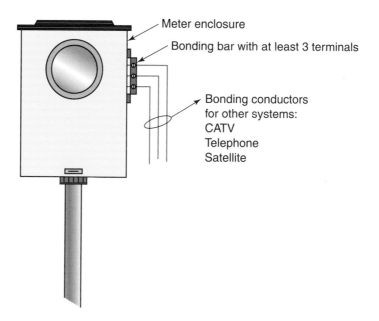

FIGURE 14-4 Intersystem bonding termination

Bonding for other systems

For bonding of other systems—typically telephone, satellite, and cable television systems—to the building grounding system, the code requires installation of an *intersystem bonding termination*, a device that provides a means for connecting grounding and bonding conductors of communications systems near the building service equipment. The bonding termination must be accessible and must have provisions for connection of at least three bonding conductors. A set of listed terminals mounted directly to the meter cabinet satisfies these requirements and may be the most common installation. Other approved methods for intersystem bonding include the installation of a bonding bar near the service enclosure, meter cabinet, or service raceway (Figure 14-4). [Ref. E3609.3]

BRANCH CIRCUITS

The IRC specifies the minimum required branch circuits, circuit ratings, conductor size, and overcurrent protection. The total number of branch circuits is determined from the total calculated load.

Branch circuit ratings and conductor size

Branch circuits must be rated according to the maximum allowable ampere rating or setting of the overcurrent protection device, typically a circuit breaker. Branch circuit requirements are summarized in Table 14-3. [Ref. E3702, Table E3702.13]

Conductor sizing

Table 14-3 contains the basic conductor sizing criteria applicable to the majority of wiring in a dwelling. For example, Type NM (nonmetallic) cable with 12 AWG copper conductors is typically satisfactory for all

TABLE 14-3 Branch circuit requirements—summary

Conductors	Circuit Rating		
	15 amp	20 amp	30 amp
Min. size (AWG) circuit conductors (copper)	14	12	10
Overcurrent-protection device: max. amp rating	15	20	30
Duplex or multiple outlet receptacle rating (amps)	15 max.	15 or 20	30
Single receptacle outlet minimum rating (amps)	15	20	30
Max. load (amps)	15	20	30

20-amp branch circuits. The code also provides ampacity tables for all wire sizes based on the material and insulation type of the conductor. When sizing wires using the ampacity tables, typically for circuits exceeding 30 amps, a number of variables must be considered, including the required temperature rating of the conductor insulation, derating for bundled conductors, and the temperature rating of the terminal. [Ref. E3705]

Overcurrent protection required

An overcurrent device, either a circuit breaker or fuse, is required to protect all ungrounded branch circuit and feeder conductors. Overcurrent protective device ratings or settings cannot exceed the allowable ampacity of the conductor. When the current passing through the conductor exceeds the ampacity rating due to overload, short circuit, or ground fault, the overcurrent device opens the circuit before heat buildup can cause damage or a fire (Table 14-4). [Ref. E3705.5, Table E3705.5.3]

Location of overcurrent devices

Overcurrent devices are installed at the point where the branch circuit conductors receive their supply, typically at the service panel. Such devices must be readily accessible and not subject to damage. The code prohibits overcurrent devices and panelboards in clothes closets or bathrooms, or located above a step. [Ref. E3705.7, E3405.4]

TABLE 14-4 Overcurrent protection rating

Copper		Aluminum or Copper-Clad Aluminum	
Size (AWG)	Max. Overcurrent Protection Device Rating (amps)	Size (AWG)	Max. Overcurrent Protection Device Rating (amps)
14	15	12	15
12	20	10	25
10	30	8	30

Note: The maximum overcurrent protection device rating shall not exceed the conductor's allowable ampacity after applying any correction or adjustment factors.

TABLE 14-5 Required branch circuits

Area or Appliance Served	Circuit(s)	Circuit May Also Serve	Alternate
Central heating equipment	Individual branch circuit sized for equipment	–	–
Kitchen countertop receptacles	Min. two 20-amp small appliance circuits	Other receptacle outlets in the kitchen, pantry, breakfast, and dining areas	Refrigerator may be served by an individual 15-amp circuit
All laundry area receptacles	Separate 20-amp circuit	–	–
Bathroom receptacle outlet(s)	Min. one 20-amp circuit	Receptacle outlets in multiple bathrooms	A circuit dedicated to a single bathroom may supply lights, fan, and other equipment in that bathroom

Required branch circuits

The minimum required branch circuits are summarized in Table 14-5. [Ref. E3703]

WIRE AND TERMINAL IDENTIFICATION

In general, electrical devices common to residential construction, such as receptacles, that connect to both the ungrounded (hot) and grounded (neutral) conductors require properly identified terminals. The identification requirement does not apply to panelboards or devices rated over 30 amps. [Ref. E3407]

Grounded (neutral) conductors and terminals

Grounded conductors not more than 6 AWG are identified by white or gray insulation or have three white stripes on other than green insulated wire. Grounded conductors greater than 6 AWG may have the same markings, but the code also permits other colors of insulation, typically black, provided the electrician applies white or gray tape or marking around the conductor at the terminal ends at the time of installation. The grounded conductor terminal on a receptacle must be substantially white in color or identified by the word "white" or letter "W." In practice, grounded conductor terminals are typically silver-colored metal (designating white), and ungrounded terminals are a darker, brass-colored metal. [Ref. E3407.1 and E3407.4]

Grounding conductors

Grounding conductors have green insulation or yellow stripes on green insulation, or are bare wires. For grounding conductors greater than 6 AWG, the code permits insulation of another color, typically black, provided the electrician applies green tape or marking around the conductor at the terminal ends and access points at the time of installation. As an alternative, the insulation may be stripped to bare wire for the entire exposed length. [Ref. E3407.2]

Ungrounded (hot) conductors

Ungrounded conductors generally must have insulation of any color other than white, gray, or green. [Ref. E3407.3]

WIRING METHODS

The IRC recognizes a number of approved wiring methods, including armored cable, metal-clad cable, and individual conductors installed in various types of metallic and nonmetallic conduit. Discussion in this chapter will focus on above-ground wiring methods using Type NM nonmetallic sheathed cable common to construction of one- and two-family dwellings and townhouses. In all cases, the cable must be approved for the location. For example, Type NM cable is not permitted underground, may not be used in wet or damp locations, and is not to be embedded in concrete. [Ref. E3801.4 and Table E3801.4]

Type NM cable

The installation requirements for Type NM cable are mainly concerned with protection from physical damage, including penetration from fasteners. The required setbacks from the edges of framing members and minimum cable support requirements are shown in Table 14-6. In unfinished basements, Type NM cable smaller than 8 AWG is prohibited from being attached directly to the bottom of joists (Figures 14-5 and 14-6). [Ref. E3802 and Table E3802.1]

TABLE 14-6 Summary of installation requirements for type NM* cable

Installation	Physical Protection	
	Minimum Setback from Edge of Framing (in.)	Physical Protection If Minimum Distance Is Not Met
Cable run parallel with the framing member or furring strip	1¼	0.0625-inch steel plate or sleeve
Cable in bored holes or notches in framing members	1¼	0.0625-inch steel plate or sleeve
Cable installed in grooves and covered	1¼ free space	0.0625-inch steel plate or sleeve
Support of Cable		
Maximum allowable on center support spacing	4.5 feet	
Maximum support distance from metal box with cable clamp	12 inches	
Maximum support distance from plastic box without cable clamp	8 inches	
Flat cables shall not be stapled on edge	—	
Conductor Length		
Minimum free conductor length at each box	6 inches with sheath removed and 3 inches outside of box	

*NM = nonmetallic sheathed cable.

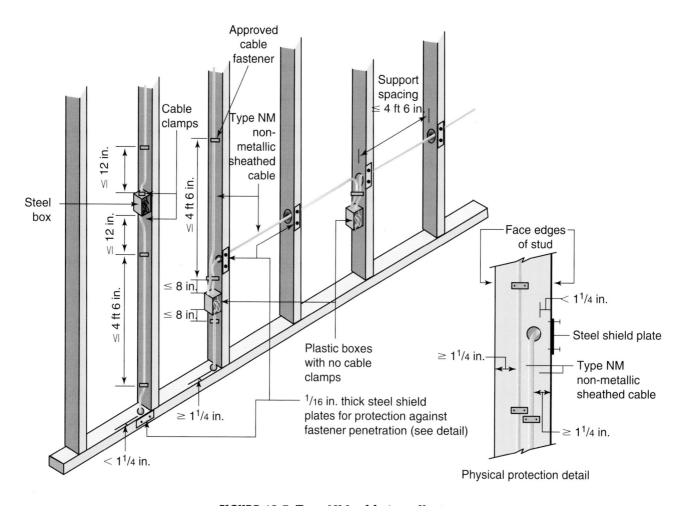

FIGURE 14-5 Type NM cable installation

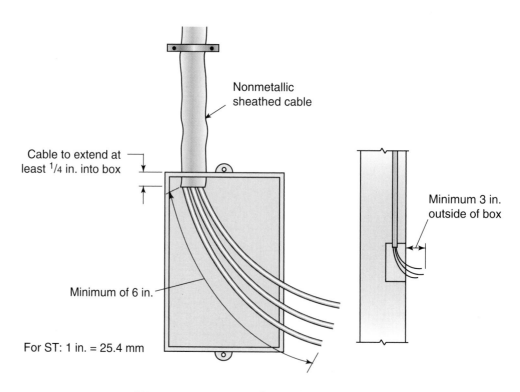

FIGURE 14-6 Free conductor length at boxes

Underground installation requirements

Cable installed underground must be listed and labeled for the location. In addition, direct burial conductors and cables emerging from the ground must be protected from 18 inches below grade to 8 feet above grade using the prescribed methods of the code. Splices underground are permitted only with specific approved methods and listed materials (Table 14-7). [Ref. E3803]

Boxes

A box is required at each splice, junction, outlet, switch, and pull point. All metal boxes require grounding by approved means, typically a short bonding jumper between a bonding screw on the metal box and the equipment grounding conductor. Each box is limited to a maximum box fill capacity based on the size of the box, the number and size of conductors entering the box, and the devices and fittings installed. Unused cable openings, typically created by the removal of a knockout plate, must be closed by approved methods to provide protection equivalent to the wall of the box. The code places limitations on boxes used for supporting ceiling-suspended paddle fans and luminaires (lighting fixtures) (Figure 14-7). [Ref. E3905, E3906.4, and E4101.6]

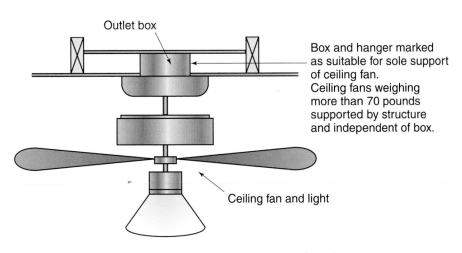

Outlet box

Box and hanger marked as suitable for sole support of ceiling fan.
Ceiling fans weighing more than 70 pounds supported by structure and independent of box.

Ceiling fan and light

FIGURE 14-7 Box supporting ceiling fan

Code Basics

Requirements for boxes supporting ceiling paddle fans:

- Marked by manufacturer as suitable
- Marked for maximum weight when supporting 36- to 70-pound fan
- Not permitted for fans over 70 pounds

Requirements for boxes supporting luminaires (lighting fixtures):

- Designed for the purpose
- Capable of supporting 50 pounds
- Permitted for luminaires weighing 50 pounds or less
- Not permitted for luminaires weighing over 50 pounds unless the box is listed and marked for the weight ●

TABLE 14-7 Minimum cover requirements, burial (in.)

Location of Wiring Method or Circuit	Type of Wiring Method or Circuit		
	Direct Burial Cables or Conductors	Rigid Metal Conduit or Intermediate Metal Conduit	Nonmetallic Raceways Listed for Direct Burial Without Concrete Encasement or Other Approved Raceways
All locations not specified below	24	6	18
Under a building	0 (in raceway only)	0	0
Under streets, highways, roads, alleys, driveways, and parking lots	24	24	24
One- and two-family dwelling driveways and outdoor parking areas, and used only for dwelling-related purposes	18	18	18

Grounding

Metal enclosures, devices, and equipment require grounding in accordance with the code. Equipment grounding conductors provide a path to ground through the equipment grounding and grounding electrode system. Type NM cable requires an insulated or bare grounding conductor used for no other purpose. [Ref. E3908]

POWER DISTRIBUTION

The IRC establishes location requirements for receptacle and lighting outlets.

Receptacle outlet locations

In addition to providing convenience for the occupants, proper placement of receptacle outlets to serve fixtures and appliances reduces electrical hazards in the home, such as may occur with the use of extension cords (Table 14-8 and Figures 14-8 through 14-11). [Ref. E3901]

TABLE 14-8 Receptacle outlet locations

Description	Maximum Spacing (ft.)	Minimum number of receptacle outlets	Location	Note
Habitable rooms	12		within 6 ft. of a door and within 6 ft. of any point along a wall	for wall space ≥ 24 in. wide; wall receptacles ≤ 5.5 ft. above the floor; floor receptacles ≤ 18 in. from wall
Kitchen wall counters	4		within 2 ft. of any point along the wall line and ≤ 20 in. above counter	for wall counter ≥ 12 in. wide
Kitchen island or peninsular counter		1	≤ 20 in. above counter	for counter ≥ 12 in. x 24 in.
Hallway		1		for hallways ≥ 10 ft.
Bathroom		1	≤ 36 in. from lavatory	measured from the outside edge of each lavatory basin
Outdoors		1 in front & 1 in back	≤ 6 feet 6 inches above grade	where there is access to grade
Deck, balcony, or porch		1	≤ 6 feet 6 inches above floor	applies to all decks, balconies, or porches ≥ 20 sq. ft. and accessible from inside dwelling
Laundry Areas		1	≤ 6 ft. from appliance	-
Basement		1	-	for basement with finished habitable space, one outlet for each separate unfinished area
Garage		1	-	-
HVAC Equipment		1	≤ 25 ft. from equipment	-

HVAC = Heating, ventilation, air conditioning

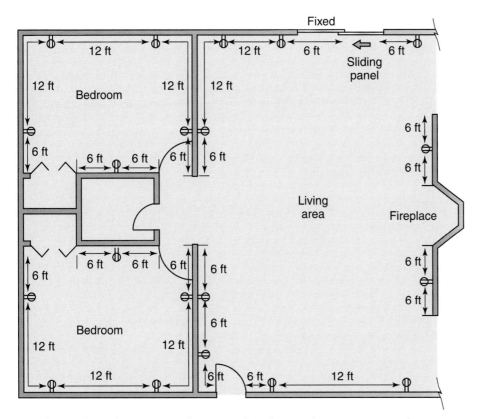

Dimensions shown are maximum spacing of general purpose receptacles

FIGURE 14-8 Habitable room receptacle outlet locations

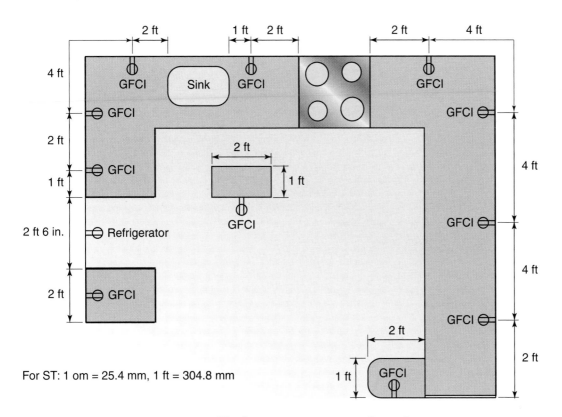

For ST: 1 om = 25.4 mm, 1 ft = 304.8 mm

FIGURE 14-9 Kitchen counter receptacle outlets

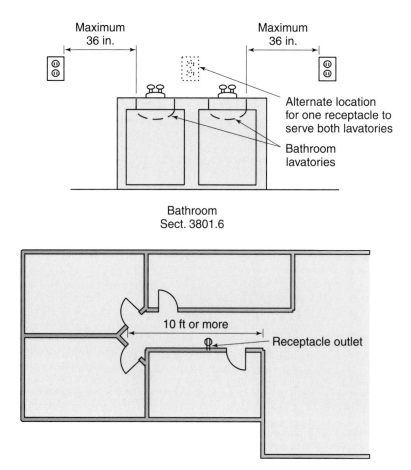

FIGURE 14-10 Bathroom and hallway receptacle outlets

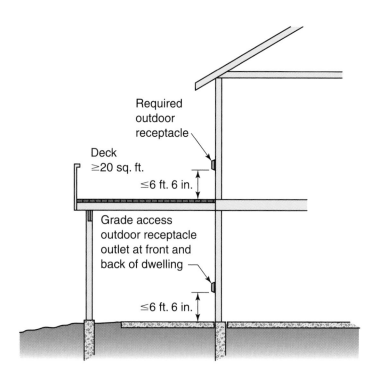

FIGURE 14-11 Outdoor receptacle outlets

Lighting outlets

Wall switch–controlled lighting outlets are required in all habitable rooms, bathrooms, hallways, storage areas and garages, at each stairway, and outside each exterior door (Figure 14-12). [Ref. E3903]

Ground-fault circuit interrupter (GFCI) protection

GFCI devices protect people from shock hazards by de-energizing a circuit or receptacle when a fault current to ground is detected. Receptacle outlets of 125 volts, 15 and 20 amps, require GFCI protection at the locations shown in Figure 14-13. [Ref. E3902]

Arc-fault circuit interrupter (AFCI) protection

AFCI devices detect unwanted arcing in the wiring of the branch circuit and open the circuit before excessive heat buildup can cause a fire. The AFCI device typically is installed in the service panel or subpanel in order to protect the entire branch circuit. The code only permits combination type AFCI devices, which are tested and listed for both branch/feeder and outlet circuit protection. AFCI protection is required for branch circuits serving 120-volt, single-phase, 15- and 20-ampere receptacle, lighting and smoke alarm outlets in most living areas of a dwelling unit, including hallways and closets. Kitchens, bathrooms, unfinished basements, garages, and outdoor outlets do not require AFCI protection. [Ref. E3902.11]

Code Basics

AFCI Protection

Branch circuits:
- 120-volt, 15- and 20-amp outlets:
  - receptacles
  - lighting
  - smoke alarms

	AFCI Required	Not Required
Bedroom	✔	
Living	✔	
Dining	✔	
Family	✔	
Recreation	✔	
Kitchen		✔
Other habitable rooms	✔	
Closet	✔	
Hallway	✔	
Bathroom		✔
Outdoors		✔
Deck, balcony, or porch		✔
Laundry		✔
Unfinished basement		✔
Garage		✔

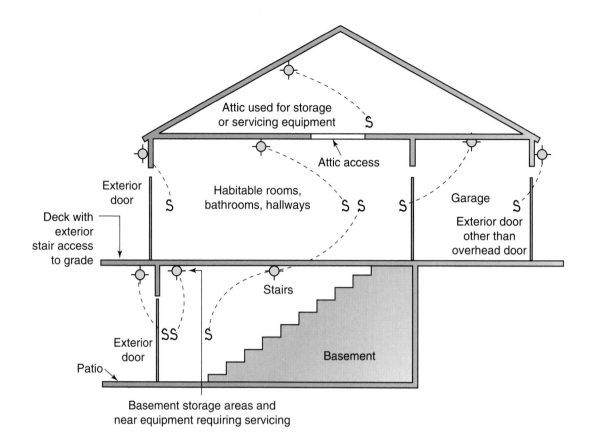

FIGURE 14-12 Wall switch–controlled lighting outlets

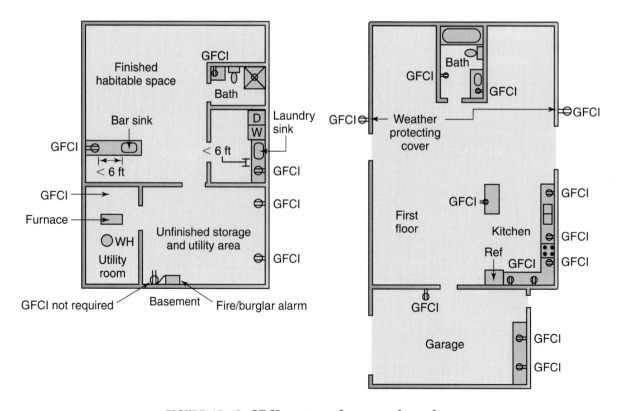

FIGURE 14-13 GFCI-protected receptacle outlets

RECEPTACLES AND LUMINAIRES

The IRC regulates the location and installation of receptacles and luminaires (lighting fixtures) to protect against fire and shock hazards.

Receptacles

A single receptacle installed on an individual branch circuit must have an ampere rating not less than that of the branch circuit. Receptacles connected to a branch circuit supplying two or more receptacles require an ampere rating as specified in Table 14-3. When installed in a wet location, 15- and 20-amp receptacles require enclosures that are weatherproof when a cord is plugged in (Figure 14-14). The code prohibits receptacles within or directly over a bathtub or shower space. [Ref. E4002]

Tamper resistant receptacles

Tamper resistant receptacles are designed to prevent the insertion of any small object, such as a paper clip, into one side of the receptacle. Both blades of an attachment plug must be inserted simultaneously to open the protective shield and allow connection to electricity. The code requires tamper resistant receptacles for all 125-volt, 15- and 20-ampere receptacles installed in dwelling units, on the outside of dwelling units, and in attached and detached garages. This added safeguard in the electrical provisions intends to reduce the number of electrical shock injuries to children (Figure 14-15). [Ref. E4002.14]

Luminaires

"Luminaire" has replaced the term "lighting fixture" in the code. The code regulates luminaires and other electrical fixtures for type, location, and clearance to combustibles.

Recessed luminaire installation and clearance

Type IC (insulation contact) recessed luminaires are listed for installation in contact with insulation and combustible materials (see Chapter 15). Other types of recessed luminaires must maintain minimum distances to combustibles and thermal insulation. [Ref. E4004.8]

Bathtub and shower areas

Specific requirements apply for fixture installations in a zone measured 3 feet horizontally and 8 feet vertically from the top of a bathtub rim or shower stall threshold (Figure 14-16). [Ref. E4003.11]

Luminaires in clothes closets

To reduce fire hazard, the code restricts the type of luminaires in clothes closets and sets minimum clearances to the storage areas. Specified clearances are measured from the fixture to the nearest point of the defined storage space (Table 14-9 and Figure 14-17). [Ref. E4003.12]

FIGURE 14-14 Weatherproof cover

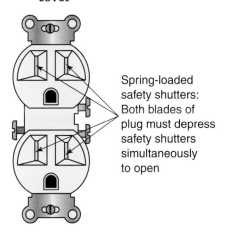

Spring-loaded safety shutters: Both blades of plug must depress safety shutters simultaneously to open

FIGURE 14-15 Tamper resistant receptacles

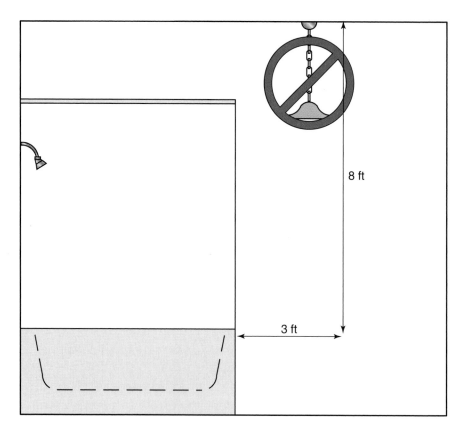

FIGURE 14-16 Luminaires in bathtub and shower areas

TABLE 14-9 Luminaires in clothes closets

Permitted Luminaire	Min. Clearance to Storage Area (in.)	Prohibited Luminaires
Incandescent		• incandescent luminaires with open or partially enclosed lamps
Surface-mounted incandescent with completely enclosed lamps	12	• pendant luminaires
Recessed incandescent with completely enclosed lamps	6	• lamp holders
Fluorescent		
Surface-mounted fluorescent	6	
Recessed fluorescent	6	
Surface-mounted fluorescent identified for use in storage space	0*	
LED		
Surface-mounted LED with completely enclosed light source	12	
Recessed LED with completely enclosed light source	6	
Surface-mounted LED identified for use in storage space	0*	

* Luminaire permitted within defined storage space

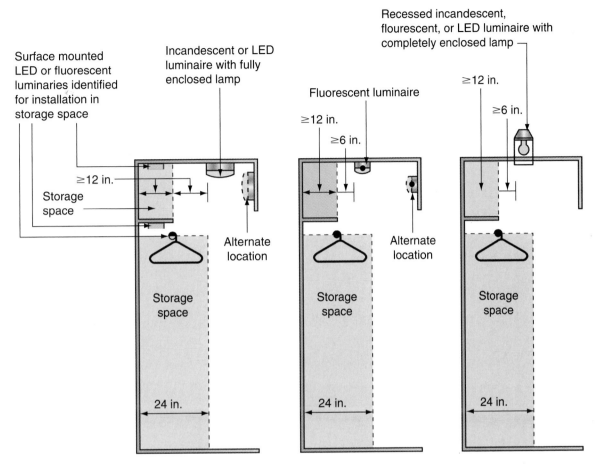

FIGURE 14-17 Luminaires in clothes closets

PART VII

Energy Conservation

Chapter 15: Energy Efficiency

Energy Efficiency

This chapter addresses the prescriptive methods of the *International Residential Code* (IRC) for effective use and conservation of energy through proper design and construction of dwellings. In addition to setting minimum requirements for the building thermal envelope enclosing conditioned space, including insulation, windows, and doors, the code regulates the sealing of penetrations to reduce air infiltration and the insulation and sealing of ductwork for heating, ventilation, and air conditioning (HVAC). Climate zones assigned to geographic location are the basis for specific thermal envelope requirements. The IRC lists the climate zone designation for each county or state corresponding to the climate zone map (Figure 15-1). As an alternative to the IRC provisions, buildings may be designed and constructed to comply with the performance requirements of the *International Energy Conservation Code* (IECC).

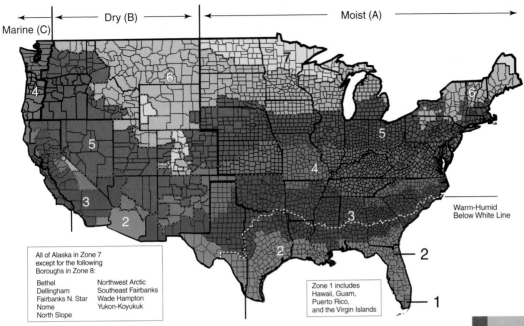

FIGURE 15-1 Climate zone map

BUILDING INSULATION

The IRC prescribes methods for identifying components of the thermal envelope to verify compliance with the energy efficiency provisions of the code. The prescriptive tabular values for minimum insulation R-values for each building component are based on the climate zone.

Insulation identification and verification

Each piece of insulation 12 inches or more in width requires a manufacturer's mark visible after installation that identifies its R-value. As an alternative, the code requires the installer to provide certification stating the type, manufacturer, and R-value of the insulation. In addition, for fiberglass or cellulose blown-in or sprayed insulation, the certification must include:

- Initial installed thickness
- Settled thickness
- Settled R-value
- Installed density
- Coverage area
- Number of bags installed

The insulation installer must sign, date, and post the certificate in a conspicuous location. This certification is in addition to the permanent certificate at the electrical panel discussed later in this chapter.

When fiberglass or cellulose insulation is blown or sprayed in the joist or truss spaces of the attic, the IRC requires fixed markers to indicate the installed thickness of the insulation. At least one marker must be installed for every 300 square feet. Minimum 1-inch-high numbers must be visible from the attic access for inspection purposes (Figure 15-2). [Ref. 1101.4]

You Should Know

R-value

R-value is used to rate the relative thermal resistance to heat flow through insulation. A higher R-value indicates greater resistance and more effective insulation. The type of insulating material, its density, and the installed thickness determine the R-value of thermal insulation. The insulation R-value does not in itself indicate the overall wall, floor, or ceiling R-value. The effectiveness of these insulated assemblies depends in part on the method of installation. For example, insulation installed between studs does not improve the resistance to heat flow through those studs. In this case, the overall average wall R-value will be less than the cavity insulation R-value. ●

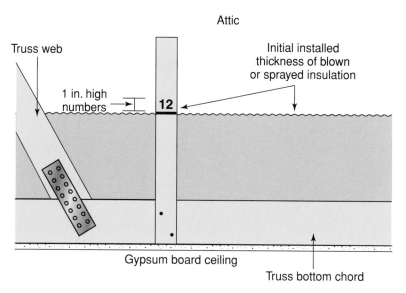

FIGURE 15-2 Attic insulation depth markers

Insulation requirements

Minimum R-values for each insulated component of the thermal envelope are shown in Table 15-1. There are a number of exceptions to the tabular values, as indicated in Table 15-2. For example, the code recognizes the increased efficiency of an energy-type or raised-heel roof truss in cold climates and reduces the required ceiling insulation R-value accordingly (Figure 15-3). The code also recognizes the practical difficulties in achieving high R-values when the rafter or joist space is too small to accommodate the required thickness of insulation (Figure 15-4). The values in Tables 15-1 and 15-2 apply to conventional wood frame construction. Due to steel's high thermal conductivity, the IRC requires higher insulation R-values and continuous insulation sheathing to provide a thermal break for steel-frame floor, wall, and ceiling construction. **[Ref. N1102.1 and N1102.2]**

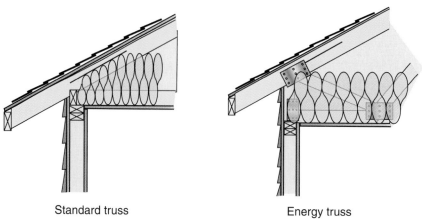

Standard truss   Energy truss

FIGURE 15-3 Insulation of standard and energy-type roof trusses

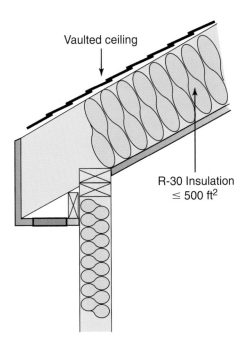

FIGURE 15-4 Reduced R-values for vaulted ceilings

TABLE 15-1 Insulation minimum R-value requirements by component

Climate Zone	Ceiling R-value	Wood Frame Wall R-value	Mass Wall R-value		Floor R-value	Basement and Crawl Space Wall R-value		Unheated Slab	
			≤50% on interior	>50% on interior		Continuous	Cavity	R-value	Depth (ft.)
1	30	13	3	4	13	0	0	0	
2	30	13	4	6	13	0	0	0	
3	30	13	5	8	19	5	13	0	
4 except marine	38	13	5	10	19	10	13	10	2
5 and marine 4	38	20	13	17	30	10	13	10	2
6	49	20	15	19	30	10	13	10	4
7 and 8	49	21	19	21	30	10	13	10	4

Slab-on-grade floors

At the perimeter of the thermal envelope, slabs with a floor surface less than 12 inches below grade require insulation with the specified R-value and minimum depth shown in Table 15-1. Heated slabs require an additional R-5 of insulation. Perimeter rigid insulation placed horizontally beneath a slab performs the same function as vertical insulation placed alongside the foundation. Therefore, the code permits the depth requirement to be satisfied by a combination of vertical and horizontal insulation. Because termites will readily excavate through foam plastic to create a path to wood construction, geographic locations subject to very heavy termite infestation are exempt from the slab-edge insulation requirements (Figure 15-5). **[Ref. N1102.2.8]**

TABLE 15-2 Alternative insulation R-value requirements by component

Climate Zone	Ceiling w/Attic Space R-value		Ceiling w/o Attic Space R-value		Wood Frame Wall w/Both Cavity and Continuous Insulation R-value		Floor R-value	
	Standard Truss or Rafters	Energy-Type Truss	General Rule	Max. 500 sf Ceiling Area	Cavity Insulation	Insulation Sheathing	General Rule	Insulation Filling Cavity
1	30	30	30	–	–	–	13	13
2	30	30	30	–	–	–	13	13
3	30	30	30	–	–	–	19	19
4 except marine	38	30	38	30	–	–	19	19
5 and marine 4	38	30	38	30	13	5	30	19
6	49	38	49	30	13	5	30	19
7 and 8	49	38	49	30	–	–	30	19

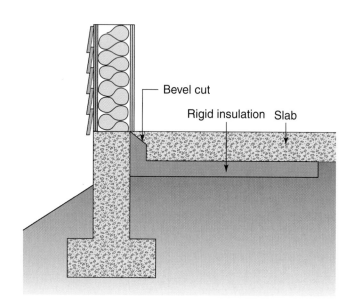

FIGURE 15-5 Slab-on-grade insulation

Crawl space walls

The IRC provides two options for the insulation of crawl spaces (referred to as underfloor spaces in other parts of the code): insulation of the floor above the crawl space or, when the crawl space is not ventilated to the outside, insulation of the exterior walls. The IRC prescribes the R-value and depth of wall insulation, as shown in Table 15-1, and specifies requirements for a vapor retarder on exposed earth of unventilated crawl spaces (Figure 15-6). See Chapter 5 of this publication for underfloor space ventilation and access requirements. **[Ref. N1102.2.9]**

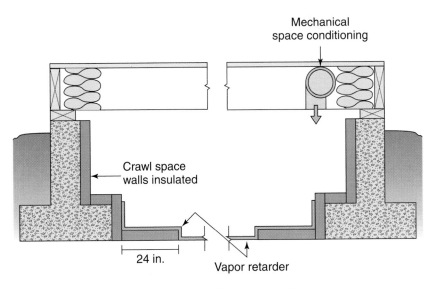

FIGURE 15-6 Crawl space insulation

WINDOWS AND DOORS

Windows, doors, and skylights are referred to as fenestration products and must meet the prescribed energy efficiency requirements. The fenestration requirements apply to both glazed doors and opaque doors. The values of the U-factor and solar heat gain coefficient (SHGC) for fenestration products are determined by the referenced standards of the National Fenestration Rating Council (NFRC). For products that lack a label certifying conformance to the applicable NFRC standard, the IRC assigns default values based on the construction of the product. The values in Table 15-3 are maximum U-factors and SHGCs for each component. In addition, the IRC regulates air infiltration rates of fenestration products, which are also measured and labeled in accordance with NFRC standards. The maximum infiltration rate for windows, skylights, and sliding glass doors is 0.3 cubic feet per minute (cfm) per square foot. For swinging doors, the maximum value is increased to 0.5 cfm per square foot (Figure 15-7). **[Ref. N1102.3 and N1102.4.4]**

TABLE 15-3 Fenestration requirements by component

Climate Zone	Fenestration U-factor	Skylight U-factor	Glazed Fenestration SHGC
1	1.2	0.75	0.35
2	0.65	0.75	0.35
3	0.50	0.65	0.35
4 except marine	0.35	0.60	–
5 and marine 4	0.35	0.60	–
6	0.35	0.60	–
7 and 8	0.35	0.60	–

FIGURE 15-7 Window NFRC energy performance label

SEALING AGAINST AIR LEAKAGE AND INFILTRATION

The building thermal envelope must be sealed at the following locations:

- All joints, seams, and penetrations, including utility penetrations
- Openings around window and door assemblies
- Knee walls and dropped ceilings
- Walls and ceilings separating the garage from conditioned spaces
- Behind tubs and showers on exterior walls
- Common walls between dwelling units
- Attic access openings
- Rim joist junction
- Other sources of infiltration

Recessed luminaires (light fixtures) can be a major source of heat loss and moisture introduced into attic spaces. Often referred to as "can lights," recessed luminaires installed in the thermal envelope, typically an insulated ceiling, require a tight seal to limit air leakage. The IRC requires an IC (insulation contact) fixture tested and labeled to conform to the specified air movement standard (Figures 15-8 and 15-9). [Ref. N1102.4]

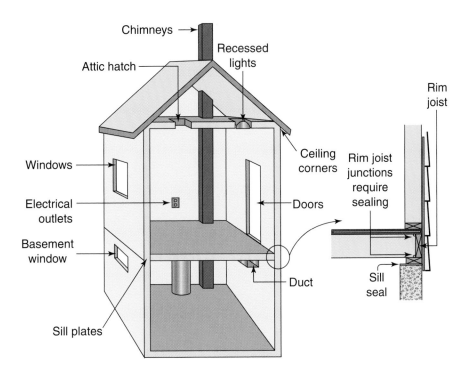

FIGURE 15-8 Typical sources of air leakage

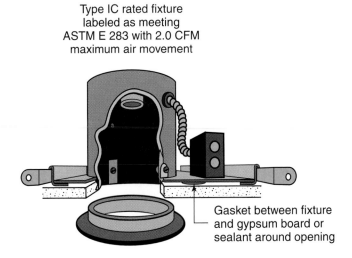

FIGURE 15-9 Recessed luminaires in thermal envelop

DUCT INSULATION AND SEALING

Supply and return ducts require at least R-8 insulation. This may be reduced to R-6 for ducts in floor trusses. Ducts entirely within conditioned spaces do not require insulation. The code requires sealing of all ducts, including framing spaces used for return air ducts. [Ref. N1103.2]

ENERGY CERTIFICATE

The IRC requires the builder or registered design professional to complete an energy efficiency certificate, listing the installed insulation and fenestration values. The certificate must also list the type and efficiency of installed heating, cooling, and water heating equipment. Because electric furnaces, baseboard heaters, and unvented gas-fired heaters may not provide the lowest energy consumption when compared to other methods of comfort heating and their energy efficiency ratings may be misleading, the IRC requires such appliances to be individually listed on the certificate without an efficiency designation. The permanent certificate is affixed to the electrical service panel, but cannot cover the service directory or other required information governed by the electrical code (Figure 15-10). [Ref. N1101.9]

Energy Efficiency Certificate		
Insulation Rating		*R*-Value
Ceiling/roof		
Walls	Frame	
	Mass	
	Basement	
	Crawl space	
Floors	Over unconditioned space	
	Slab edge	
Ducts	Outside conditioned spaces	
Glass and Door Rating	NFRC U-Factor	NFRC SHGC
Window		
Opaque door		
Skylight		
Heating and Cooling Equipment	Type	Efficiency
Heating system		AFUE
Cooling system		SEER
Water heater		EF

Indicate if the following have been installed (an efficiency shall not be listed):

☐ electric furnace　☐ gas-fired unvented room heater　☐ baseboard electric heater

Designer _____

Builder _____

Date _____

FIGURE 15-10 Permanent energy certificate

PART

VIII

Protection from Other Hazards

Chapter 16: Other Hazards of the Built Environment

CHAPTER 6

Other Hazards of the Built Environment

This chapter provides information on structural and environmental hazards often associated with dwelling and accessory building construction. Other than termite protection, the topics covered are generally not regulated in the body of the *International Residential Code* (IRC). Requirements for swimming pools and radon control appear in the IRC appendices and are not in force unless specifically adopted by the jurisdiction. Lead and asbestos hazards are encountered primarily in building remodeling or rehabilitation and are usually governed by state regulations.

TERMITE PROTECTION

Subterranean termites cause significant damage to concealed structural and nonstructural wood components. They thrive in moist ground and usually invade homes by building mud tunnels on the surface of foundations. They may also travel inside hollow block masonry, through plastic foam insulation, or directly through untreated wood in contact with the ground or organic materials such as mulch. Mud tunnels may be observed only when the minimum clearances above grade are maintained as required by the IRC. Termite activity may also be deterred by blocking their access points above the foundation with termite shields and pressure-preservative-treated foundation plates.

The termite protection provisions of the IRC are applicable to all geographic areas subject to termite damage. When adopting the IRC, the jurisdiction designates the level of hazard based on a history of local subterranean termite damage. The infestation probability map in Figure 16-1 shows the approximate areas of hazard rated from none or slight to very heavy. [Ref. R301.2 and Table R301.2(1)]

Termite control methods

For wood frame construction, the IRC requires one or more of five methods of protection in all areas subject to termite damage: chemical termiticide treatment, termite baiting, pressure-preservative-treated wood, naturally termite-resistant wood, and physical barriers. Liquid termiticide treatment is the most common termite control method. These chemicals are placed in the ground to kill or repel termites. There are also formulations available to treat wood on-site. Baiting systems are used when

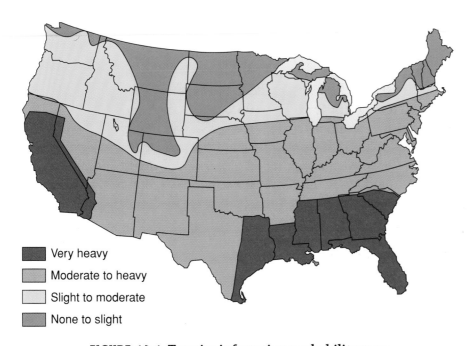

Very heavy
Moderate to heavy
Slight to moderate
None to slight

FIGURE 16-1 Termite infestation probability map

application of a liquid termiticide is not feasible or practical. The IRC requires pressure-preservative-treated wood or naturally durable wood in the locations described in Chapter 6 of this publication for protection against decay and termites. Physical barriers are typically continuous corrosion-resistant metal termite shields placed at the top of foundation walls or piers. When installed on exterior foundation walls, termite shields have limited effectiveness and require one or more additional approved methods. **[Ref. R318.1 and R318.3]**

Foam plastic protection

In geographic locations where the termite probability designation is "very heavy," the IRC generally prohibits unprotected foam plastic insulation on the exterior surface of foundation walls or under slab foundations on grade. **[Ref. R316.7 and R318.4]**

SWIMMING POOLS AND HOT TUBS

IRC Appendix Chapter G establishes minimum requirements for protection against physical injury and drowning associated with swimming pools on lots of one- and two-family dwellings. The barrier requirements intend to prevent unattended small children from entering the pool area, thereby reducing the risk of drowning. Entrapment protection is necessary to prevent serious injury or drowning associated with pump suction.

Barriers

The barrier provisions apply to outdoor and indoor swimming pools and hot tubs with water depths greater than 24 inches. An approved safety cover for a hot tub satisfies the barrier requirement. Required pool barriers or fences must be at least 48 inches high and constructed to prevent climbing by children. Access gates must be self-closing and self-latching with the hardware arranged so the gate cannot be opened from the side opposite the pool. For above-ground pools, the wall of the pool may serve as the barrier, provided it meets the 48-inch height requirement and has either a removable ladder or steps (Figures 16-2 and 16-3). **[Ref. AG105.2 Item 9 and AG105.3]**

In addition to outside walls or fences as barriers, the appendix chapter requires protection for occupants of the residence as well. Unless the pool is equipped with an approved power safety cover, doors with direct access to an indoor or outdoor swimming pool require an audible warning alarm system. The alarm must be audible in all portions of the house and is activated when the door is opened. The code provides for a temporary deactivation device located out of reach of a young child. The alarm may be deactivated for a maximum of 15 seconds before automatically resetting. The appendix also authorizes the building official to accept alternate and equivalent self-closing and self-latching devices in lieu of the door alarm (Figure 16-4). **[Ref. AG105.2 Item 9 and AG105.3]**

Code Basics

Swimming pool:

- Contains water over 24 inches deep
- Intended for recreational swimming or bathing
- Includes in-ground or above-ground pools
- Includes hot tubs and spas ●

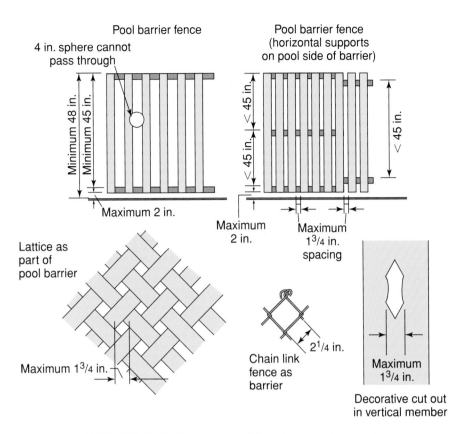

FIGURE 16-2 Swimming pool barrier requirements

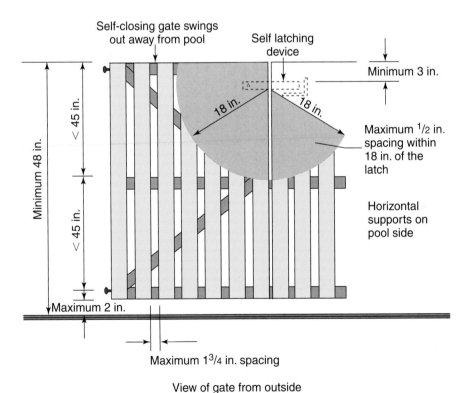

FIGURE 16-3 Pool access gate requirements

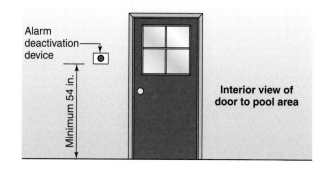

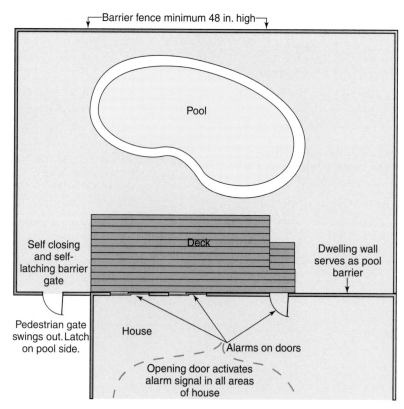

FIGURE 16-4 Pool barrier and door alarm requirements

Entrapment protection

Underwater entrapment occurs when parts of the body cover the suction outlet, creating a vacuum, or when long hair or clothing becomes entangled in a drain without a safety cover, potentially resulting in severe injury or drowning. The IRC appendix references the appropriate standard for avoidance of entrapment hazards to improve pool safety. Based on the latest information and technology to address all forms of entrapment hazards and their underlying causes, the code and the referenced standard provide that all swimming pools and spas be equipped with proper anti-entrapment drain covers and circulation and drainage systems. [Ref. AG106]

RADON CONTROL

Appendix Chapter F establishes prescriptive provisions for radon resistant new construction. The prescribed measures act to reduce introduction of soil gases potentially containing radon into the living space of a dwelling and to provide a cost-effective means of mitigation should testing reveal unacceptable levels of radon after construction. A decay product of uranium, radon is a radioactive gas that occurs naturally in the soil in varying concentrations and over time can cause damage to the lungs and increases the potential for developing lung cancer. It is not possible to predict with accuracy or certainty the levels of radon that will occur in a given house until the building is enclosed and tested. Houses in the same neighborhood, because of varying radon concentrations in the soil, may experience very different levels of radon.

The primary means for reducing entry of radon gas into the dwelling is through sub-slab depressurization. This method seals all entry points of the slab and basement foundation wall and vents the area below the slab out through the roof. Effectiveness of this system depends on an air-permeable granular base, such as sand or gravel, below the slab. Even a passive system without a fan on the vent pipe creates a chimney effect and intends to maintain a lower air pressure below the slab than occurs in the dwelling. If testing in the lowest living space of the dwelling reveals radon levels above the action level, an approved in-line fan may be installed in the vent pipe in the attic. The fan runs continuously to draw air from below the slab. With lower sub-slab pressure, soil gases will follow the path of least resistance out through the roof rather than migrating into the living space (Figures 16-5 to 16-7). [Ref. AF103]

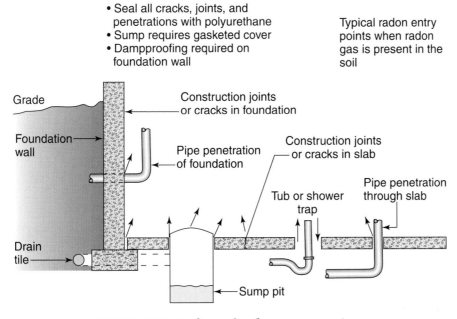

• Seal all cracks, joints, and penetrations with polyurethane
• Sump requires gasketed cover
• Dampproofing required on foundation wall

Typical radon entry points when radon gas is present in the soil

Grade

Foundation wall

Construction joints or cracks in foundation

Pipe penetration of foundation

Construction joints or cracks in slab

Tub or shower trap

Pipe penetration through slab

Drain tile

Sump pit

FIGURE 16-5 Sealing of soil gas-entry points

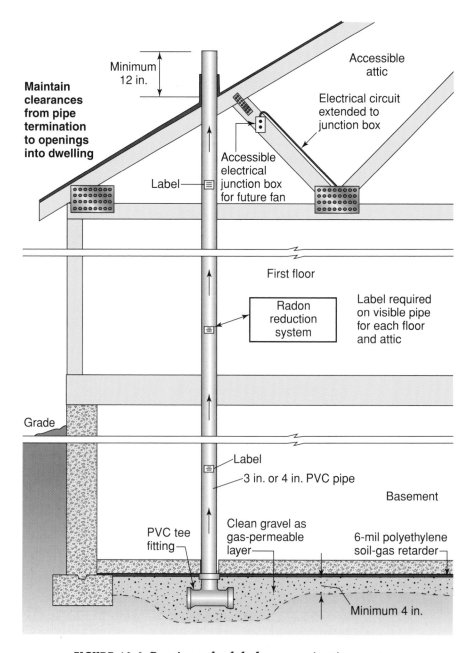

Maintain clearances from pipe termination to openings into dwelling

Minimum 12 in.

Accessible attic

Electrical circuit extended to junction box

Label

Accessible electrical junction box for future fan

First floor

Radon reduction system

Label required on visible pipe for each floor and attic

Grade

Label

3 in. or 4 in. PVC pipe

Basement

PVC tee fitting

Clean gravel as gas-permeable layer

6-mil polyethylene soil-gas retarder

Minimum 4 in.

FIGURE 16-6 Passive sub-slab depressurization system

LEAD

Lead is a toxic substance that can cause harmful health effects when ingested. Children are particularly susceptible to the effects of lead poisoning, which is most commonly caused by the ingestion of lead paint dust or chips in older houses. A secondary cause is from drinking water containing lead that has leached from lead pipes or solder.

Paint used in homes prior to 1960 typically contained relatively high levels of lead. These levels decreased until 1978, when the Consumer Product Safety Commission banned lead-based paint. Therefore, the risk of lead poisoning is greatest in houses constructed before 1960. Even so, paint surfaces maintained in good condition pose little risk. It is through

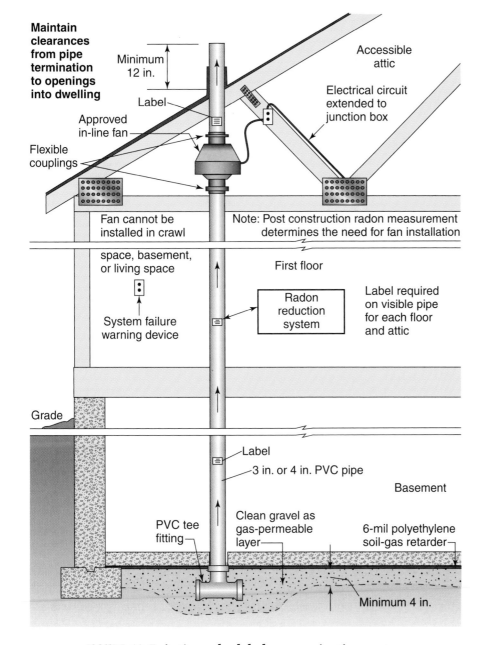

Maintain clearances from pipe termination to openings into dwelling

Minimum 12 in.

Accessible attic

Label

Approved in-line fan

Flexible couplings

Electrical circuit extended to junction box

Fan cannot be installed in crawl space, basement, or living space

Note: Post construction radon measurement determines the need for fan installation

First floor

System failure warning device

Radon reduction system

Label required on visible pipe for each floor and attic

Grade

Label

3 in. or 4 in. PVC pipe

Basement

PVC tee fitting

Clean gravel as gas-permeable layer

6-mil polyethylene soil-gas retarder

Minimum 4 in.

FIGURE 16-7 Active sub-slab depressurization system

sanding, scraping, peeling, flaking, or chipping that lead paint becomes hazardous. Care must be taken in remodeling, renovation, and maintenance to contain and properly dispose of dust and debris.

Communities often implement nuisance or housing ordinances to regulate the maintenance of buildings. The *International Property Maintenance Code* (IPMC) requires all interior and exterior surfaces, including windows, doors, and trim, to be maintained in good condition. The IPMC requires repairing, removing, or covering any peeling, chipping, flaking, or abraded paint. **[Ref. IPMC 304.2 and 305.3]**

Lead is also a concern in drinking water supplied through older piping systems of lead pipe, fittings, or solder. Such materials have very limited use in new construction. The IRC regulates lead content in systems of potable-water supply. Solders and fluxes are limited to a maximum of 0.2 percent lead. The code limits pipe and fittings used in the water supply system to a maximum of 8 percent lead, though the most common types of water piping systems are copper and plastic. [Ref. P2905.2 and P2904.13]

ASBESTOS

Asbestos fibers are a health hazard when inhaled into the lungs. Modern construction materials no longer contain asbestos, and its use is generally prohibited, even in the repair of existing buildings. For example, while IRC Appendix Chapter J, "Existing Buildings and Structures," and the *International Existing Building Code* (IEBC) generally permit repairs using similar materials, they specifically prohibit the use of asbestos and other hazardous materials. [Ref. IRC AJ301.1 and IEBC 502.2]

Asbestos is often found in the following materials in older homes:

- Asbestos blanket, paper, or tape on steam pipes, boilers, and ducts of heating, ventilation, and air conditioning (HVAC) systems
- Vinyl asbestos floor tile and other resilient flooring and adhesives
- Cement board sheets used adjacent to wood-burning stoves for protection of combustible materials
- Gypsum board, plaster, and ceiling tile
- Patching and joint compounds and spray-on texture materials
- Asbestos cement sheet siding, siding and roofing shingles, and other roofing materials

The asbestos materials listed above are not hazardous while they remain in place and intact. Drilling, sawing, sanding, or removing such products is likely to create hazardous dust and airborne asbestos fibers. Special precautions must be followed to contain and dispose of asbestos materials properly and safely during abatement, repair, or renovation. Many states have laws regulating the removal and handling of asbestos-containing materials, as well as related permitting and licensing for such operations.

GLOSSARY

A

accepted engineering practice – Engineered analysis based on well-established principles of mechanics and conforming to accepted principles, tests, or standards of nationally recognized technical authorities.

accessory structure – A building or structure accessory to and incidental to the dwelling, such as a detached garage or shed. Limited to 3000 square feet in area and two stories in height, accessory buildings must also be located on the same lot as the dwelling.

air gap – Open space between the potable-water outlet and the flood-level rim of the fixture, effectively separating the potable water from the source of contamination.

ampacity – The current, in amperes, that a conductor can carry continuously under the conditions of use without exceeding its temperature rating.

approved – Acceptable to the building official.

approved agency – An established and recognized agency regularly engaged in conducting tests or furnishing inspection services, when such agency has been approved by the building official.

arc-fault circuit interrupter (AFCI) – A device intended to provide protection from the effects of arc faults by recognizing characteristics unique to arcing and by functioning to de-energize the circuit when an arc fault is detected.

AWG – American Wire Gauge; indicates the size of the wire. As the AWG number decreases, the wire size (diameter) increases.

B

bonding – Permanent joining of metallic parts to form an electrically conductive path that will ensure electrical continuity and the capacity to conduct safely any current likely to be imposed.

bonding jumper – A reliable conductor to ensure the required electrical conductivity between metal parts required to be electrically connected.

building drain – Lowest drainage piping inside the house; extends 30 inches beyond the exterior walls to connect to the building sewer. The building drain collects the discharge from all other drainage piping in the dwelling.

building official – The officer or other designated authority charged with the administration and enforcement of the IRC.

building sewer – That part of the drainage system that extends from the end of the building drain 30 inches outside of the building and conveys its discharge to a public sewer, private sewer, individual sewage-disposal system, or other approved point of disposal.

C

category IV vented appliance – An appliance that operates with a positive vent static pressure and with a vent gas temperature that is capable of causing excessive condensate production in the vent. Category IV appliances are referred to as very high efficiency condensing appliances. They rely on mechanical means rather than gravity to vent the low-temperature flue gases to the outside. Vents must be of noncorrosive material as specified by the appliance manufacturer, such as PVC pipe.

cricket – A sloped flashing on the up-roof side of a chimney to divert water from above the chimney to each side.

D

dead loads – The actual weight of all materials of construction incorporated into the building, including but not limited to walls, floors, roofs, ceilings, stairways, built-in partitions, finishes, cladding, and fixed service equipment. Dead loads are fixed and considered permanent.

design flood elevation – The elevation subject to a 1 percent or greater chance of flooding in any year or the otherwise legally designated elevation in a flood hazard area.

d.f.u. – In plumbing, drainage fixture units. A measure of probable discharge into the drainage system by various types of plumbing fixtures, used to size DWV piping systems. The drainage fixture unit value for a particular fixture depends on its volume rate of drainage discharge, on the time duration of a single drainage operation, and on the average time between successive operations.

direct-vent appliance – Fuel-burning appliance with a sealed combustion system that draws all air for combustion from the outside atmosphere and discharges all flue gases to the outside atmosphere.

draft stop – Material or construction installed to restrict the movement of air within open spaces of concealed areas of buildings.

dwelling – A building that contains one or two dwelling units occupied or intended to be occupied for living purposes.

dwelling unit – A single unit providing complete independent living facilities for one or more persons, including permanent provisions for living, sleeping, eating, cooking, and sanitation.

DWV – The drainage, waste, and vent (DWV) system consists of all piping for conveying wastes from plumbing fixtures, appliances, and appurtenances, including fixture traps and venting systems.

E

effectively grounded – An electrical term meaning intentionally connected to earth through a ground connection or connections of sufficiently low impedance and having sufficient current-carrying capacity to prevent the buildup of voltages that may result in undue hazards to connected equipment or to persons.

equipment grounding conductor – Conductor used to connect the non–current-carrying metal parts of equipment, conduit, and other enclosures to the system grounded (neutral) conductor, the grounding electrode conductor, or both, at the service equipment.

Evaluation Service (ES) report – A report that presents the findings of ICC-ES as to the compliance with code requirements of the subject of the report—a particular building product, component, method, or material.

F

fenestration – Skylights, roof windows, vertical windows (whether fixed or moveable), opaque doors, glazed doors, glass block, and combination opaque/glazed doors.

fireblocking – Building materials installed to resist the free passage of flame to other areas of the building through concealed spaces.

fire separation distance – The distance measured from the building face to the closest interior lot line; to the centerline of a street, an alley, or public way; or to an imaginary line between two buildings on the lot. The distance is measured at a right angle from the face of the wall.

flood hazard area – The area within a flood plain subject to a 1 percent or greater chance of flooding in any year, or the area otherwise legally designated as a flood hazard area.

floodway – The channel of the river, creek, or other watercourse and the adjacent land areas that must be reserved in order to discharge the base flood without cumulatively increasing the water surface elevation more than a designated height.

G

GFCI – Ground-fault circuit interrupter. A device intended to protect people, it functions to de-energize a circuit within an established period of time when a current to ground exceeds a certain value, indicating a fault.

gpm – Gallons per minute.

grade plane – A reference plane representing the average of the finished ground level adjoining the building at all exterior walls. Where the finished ground level slopes away from the exterior walls, the reference plane shall be established by the lowest points within the area between the building and the lot line or, where the lot line is more than 6 feet from the building between the structure and a point 6 feet from the building.

ground – A conducting connection between an electrical circuit or equipment and the earth.

grounded conductor – System or circuit conductor that is intentionally grounded. This is commonly referred to as the *neutral* and is sometimes confused with the terminology for grounding conductor. Grounded conductors (wires) generally are marked with white.

grounding conductor – A conductor used to connect equipment or the grounded circuit of a wiring system to a grounding electrode or electrodes. The grounding conductor typically bonds to the grounded (neutral) conductor only at the service equipment, and typically has green markings or is bare wire.

grounding electrode – Device that establishes an electrical connection to earth. Ground rods, concrete-encased reinforcing bars, and underground copper water service piping are examples of grounding electrodes.

grounding electrode conductor – Conductor used to connect the grounding electrode(s) to the equipment grounding conductor, to the grounded conductor, or to both, typically at the service.

H

habitable attic – A finished or unfinished area, not considered a story, with an occupiable floor area at least 70 square feet complying with the ceiling height requirements and enclosed by the roof assembly above and the floor-ceiling assembly below.

habitable space – Space in a building for living, sleeping, eating, or cooking. Bathrooms, toilet rooms, closets, halls, storage or utility spaces, and similar areas are not considered habitable spaces.

I

inertia – The property of matter that produces a tendency of objects at rest to remain at rest and objects in uniform motion to remain in uniform motion.

interlayment – In the application of wood shakes, interlayment is an 18-inch-wide strip of at least No. 30 felt shingled between each course of shakes in such a manner that no felt is exposed to the weather.

J

jalousie – A window, blind, or shutter having adjustable horizontal slats. A jalousie window has overlapping glass slats which open to allow the passage of air and light.

K

kcmil – Thousands of circular mils. In the North American electrical industry, conductors larger than 4/0 AWG are generally identified by the area in thousands of circular mils (kcmil), where 1 kcmil = 0.5067 mm². An older abbreviation for 1000 circular mils is *mcm*.

L

labeled – Devices, equipment, or materials bearing a label, seal, symbol, or other identifying mark of a testing laboratory that attests to compliance with a specific standard.

light-frame construction – A type of construction whose vertical and horizontal structural elements are formed primarily by a system of repetitive wood or cold-formed steel framing members.

listed, listing – Terms referring to equipment that is listed by an approved testing agency as complying with nationally recognized standards when installed in accordance with the manufacturer's installation instructions.

live loads – Loads produced by the use and occupancy of the building, such as people and furnishings, and not including construction or environmental loads such as those for wind, snow, rain, earthquake, flood, or dead loads. Live loads are variable and temporary.

loads – Forces that are applied to the structural system of the building.

luminaire – Complete lighting unit (lighting fixture), consisting of a lamp or lamps, together with parts designed to distribute the light, to position and protect the lamps and ballast, and to connect the lamps to the power supply.

M

manufactured home – A structure, transportable in one or more sections, which in the traveling mode is 8 body feet or more in width or 40 body feet or more in length, or, when erected on site, is 320 square feet or more, and which is built on a permanent chassis and designed to be used as a dwelling with or without a permanent foundation. A mobile home is considered a manufactured home.

means of egress – The path of travel from any occupied portion of the building to the outdoors. Stairways, ramps, landings, hallways, and doors are components of the means of egress.

multipurpose fire sprinkler system – A system that supplies domestic water to both plumbing fixtures and fire sprinklers.

O

overcurrent – Any current in excess of the rated current of equipment or the ampacity of a conductor. It may result from overload, short circuit, or ground fault.

overcurrent protection device – A circuit breaker, fuse, or other device that protects the circuit by opening the device, thereby disconnecting power to the circuit, when the current reaches a value that will cause an excessive or dangerous temperature rise in conductors (overcurrent).

R

registered design professional – An individual, such as an engineer or architect, who is registered or licensed to practice a design profession as defined by the statutory requirements of the professional registration laws of the state or jurisdiction in which the project is to be constructed.

S

seismic – Characteristic of or having to do with earthquake ground motion.

stack vent – Continuation of the stack above the highest horizontal drain connection.

substantial improvement – Any repair, reconstruction, rehabilitation, addition, or improvement of a building or structure, the cost of which equals or exceeds 50 percent of the market value of the structure before the improvement or repair is started. A determination of substantial improvement relates to buildings located in a flood hazard area.

T

townhouse – A dwelling unit constructed in a group of three or more attached units in which each unit extends from foundation to roof and with open space on at least two sides.

U

ungrounded conductor – Often referred to as the "hot" conductor (wire). The insulation color is typically black or red but may be of any color other than white, green, or gray.

W

waste stack – A main line of vertical DWV piping that conveys only liquid sewage not containing fecal material.

wood structural panel – A panel manufactured from veneers or wood strands or wafers bonded together with waterproof synthetic resins or other suitable bonding systems. Examples of wood structural panels are plywood, OSB, and composite panels.

INDEX

ICC EVALUATION SERVICE

Most Widely Accepted and Trusted

Innovative Building Products

Make sure they are up to code with ICC-ES Evaluation Reports

The ICC-ES Solution

ICC Evaluation Service® (ICC-ES®), a subsidiary of ICC®, was created to assist code officials and industry professionals in verifying that new and innovative building products meet code requirements. This is done through a comprehensive evaluation process that results in the publication of ICC-ES Evaluation Reports for those products that comply with requirements in the code or acceptance critera. Today, more code officials prefer using ICC-ES Evaluation Reports over any other resource to verify products comply with codes.

FREE Access to ICC-ES Evaluation Reports!

ICC EVALUATION SERVICE
Most Widely Accepted and Trusted

ICC-ES Evaluation Report　　　　**ESR-4802**

Issued March 1, 2008

This report is subject to re-examination in one year.

www.icc-es.org | 1-800-423-6587 | (562) 699-0543　　*A Subsidiary of the International Code Council®*

DIVISION: 07—THERMAL AND MOISTURE PROTECTION
Section: 07410—Metal Roof and Wall Panels

REPORT HOLDER:

ACME CUSTOM-BILT PANELS
52380 FLOWER STREET
CHICO, MONTANA 43820
(808) 664-1512
www.custombiltpanels.com

EVALUATION SUBJECT:

CUSTOM-BILT STANDING SEAM METAL ROOF PANELS: CB-150

1.0　EVALUATION SCOPE

Compliance with the following codes:

- 2006 *International Building Code®* (IBC)
- 2006 *International Residential Code®* (IRC)

Properties evaluated:

- Weather resistance
- Fire classification
- Wind uplift resistance

2.0　USES

Custom-Bilt Standing Seam Metal Roof Panels are steel panels complying with IBC Section 1507.4 and IRC Section R905.10. The panels are recognized for use as Class A roof coverings when installed in accordance with this report.

3.0　DESCRIPTION

3.1　Roofing Panels:

Custom-Bilt standing seam roof panels are fabricated in steel and are available in the CB-150 and SL-1750 profiles. The panels are roll-formed at the jobsite to provide the standing seams between panels. See Figures 1 and 3 for panel profiles. The standing seam roof panels are roll-formed from minimum No. 24 gage [0.024 inch thick (0.61 mm)] cold-formed sheet steel. The steel conforms to ASTM A 792, with an aluminum-zinc alloy coating designation of AZ50.

3.2　Decking:

Solid or closely fitted decking must be minimum ¹⁵/₃₂-inch-thick (11.9 mm) wood structural panel or lumber sheathing, complying with IBC Section 2304.7.2 or IRC Section R803, as applicable.

4.0　INSTALLATION

4.1　General:

Installation of the Custom-Bilt Standing Seam Roof Panels must be in accordance with this report, Section 1507.4 of the IBC or Section R905.10 of the IRC, and the manufacturer's

published installation instructions. The manufacturer's installation instructions must be available at the jobsite at all times during installation. The roof panels must be installed on solid or closely fitted decking, as specified in Section 3.2. Accessories such as gutters, drip angles, fascias, ridge caps, window or gable trim, valley and hip flashings, etc., are fabricated to suit each job condition. Details must be submitted to the code official for each installation.

4.2　Roof Panel Installation:

4.2.1　CB-150: The CB-150 roof panels are installed on roof shaving a minimum slope of 2:12 (17 percent). The roof panels are installed over the optional underlayment and secured to the sheathing with the panel clip. The clips are located at each panel rib side lap spaced 6 inches (152 mm) from all ends and at a maximum of 4 feet (1.22 m) on center along the length of the rib, and fastened with a minimum of two No. 10 by 1-inch pan head corrosion-resistant screws. The panel ribs are mechanically seamed twice, each pass at 90 degrees, resulting in a double-locking fold.

4.3　Fire Classification:

The steel panels are considered Class A roof coverings in accordance with the exception to IBC Section 1505.2 and IRC Section R902.1.

4.4　Wind Uplift Resistance:

The systems described in Section 3.0 and installed in accordance with Sections 4.1 and 4.2 have an allowable wind uplift resistance of 45 pounds per square foot (2.15 kPa).

5.0　CONDITIONS OF USE

The standing seam metal roof panels described in this report comply with, or are suitable alternatives to what is specified in, those codes listed in Section 1.0 of this report, subject to the following conditions:

5.1 Installation must comply with this report, the applicable code, and the manufacturer's published installation instructions. If there is a conflict between this report and the manufacturer's published installation instructions, this report governs.

5.2 The required design wind loads must be determined for each project. Wind uplift pressure on any roof area must not exceed 45 pounds per square foot (2.15 kPa).

6.0　EVIDENCE SUBMITTED

Data in accordance with the ICC-ES Acceptance Criteria for Metal Roof Coverings (AC166), dated October 2007.

7.0　IDENTIFICATION

Each standing seam metal roof panel is identified with a label bearing the product name, the material type and gage, the Acme Custom-Bilt Panels name and address, and the evaluation report number (ESR-4802).

ICC-ES Evaluation Reports are not to be construed as representing aesthetics or any other attributes not specifically addressed, nor are they to be construed as an endorsement of the subject of the report or a recommendation for its use. There is no warranty by ICC Evaluation Service, Inc., express or implied, as to any finding or other matter in this report, or as to any product covered by the report.

© 2008 Copyright

 ANSI

Page 1 of 1

William Gregory
Building and Plumbing Inspector
Town of Yorktown, New York

"We've been using ICC-ES Evaluation Reports as a basis of product approval since 2002. I would recommend them to any jurisdiction building department, particularly in light of the many new products that regularly move into the market. It's good to have a group like ICC-ES evaluating these products with a consistent and reliable methodology that we can trust."

Becky Baker, CBO
Director/Building Official
Jefferson County, Colorado

"The ICC-ES Evaluation Reports are designed with the end user in mind to help determine if building products comply with code. The reports are easily accessible, and the information is in a format that is useable by plans examiners and inspectors as well as design professionals and contractors."

VIEW ICC-ES EVALUATION REPORTS ONLINE!

www.icc-es.org

09-02246

INTRODUCING SAVE™

Sustainable **A**ttributes **V**erification and **E**valuation™—New from ICC-ES®

The new SAVE™ program from ICC-ES® provides the most trusted third-party verification available today for sustainable construction products. Under this program, ICC-ES evaluates and confirms product's sustainable attributes. The SAVE™ program may also assist in identifying products that help qualify for points under major green rating systems such as US Green Building Council's LEED, Green Building Initiative's Green Globes or ICC/NAHB's proposed National Green Building Standard (NGBS). When it comes to making sure that products possess the sustainable attributes claimed, you can trust ICC-ES SAVE.

FOR MORE INFORMATION ABOUT SAVE: 1-800-423-6587 | www.icc-es.org/save

ICC EVALUATION SERVICE

8-61804-66

Approve plumbing products with a name you have come to trust.

ICC-ES PMG Listing Program

When it comes to approving plumbing, mechanical, or fuel gas (PMG) products, ask manufacturers for their ICC-ES PMG Listing. Our listing ensures code officials that a thorough evaluation has been done and that a product meets the requirements in both the codes and the standards. ICC-ES is the name code officials prefer when it comes to approving products.

Look for the Mark

FOR DETAILS! 1-800-423-6587, x5478 | www.icc-es.org/pmg

 **ICC EVALUATION SERVICE**

8-61804-65

Don't Miss Out On Valuable ICC Membership Benefits. Join ICC Today!

Join the largest and most respected building code and safety organization. As an official member of the International Code Council®, these great ICC® benefits are at your fingertips.

EXCLUSIVE MEMBER DISCOUNTS

ICC members enjoy exclusive discounts on codes, technical publications, seminars, plan reviews, educational materials, videos, and other products and services.

TECHNICAL SUPPORT

ICC members get expert code support services, opinions, and technical assistance from experienced engineers and architects, backed by the world's leading repository of code publications.

FREE CODE—LATEST EDITION

Most new individual members receive a free code from the latest edition of the International Codes®. New corporate and governmental members receive one set of major International Codes (Building, Residential, Fire, Fuel Gas, Mechanical, Plumbing, Private Sewage Disposal).

FREE CODE MONOGRAPHS

Code monographs and other materials on proposed International Code revisions are provided free to ICC members upon request.

PROFESSIONAL DEVELOPMENT

Receive Member Discounts for on-site training, institutes, symposiums, audio virtual seminars, and on-line training! ICC delivers educational programs that enable members to transition to the I-Codes®, interpret and enforce codes, perform plan reviews, design and build safe structures, and perform administrative functions more effectively and with greater efficiency. Members also enjoy special educational offerings that provide a forum to learn about and discuss current and emerging issues that affect the building industry.

ENHANCE YOUR CAREER

ICC keeps you current on the latest building codes, methods, and materials. Our conferences, job postings, and educational programs can also help you advance your career.

CODE NEWS

ICC members have the inside track for code news and industry updates via e-mails, newsletters, conferences, chapter meetings, networking, and the ICC website (www.iccsafe.org). Obtain code opinions, reports, adoption updates, and more. Without exception, ICC is your number one source for the very latest code and safety standards information.

MEMBER RECOGNITION

Improve your standing and prestige among your peers. ICC member cards, wall certificates, and logo decals identify your commitment to the community and to the safety of people worldwide.

ICC NETWORKING

Take advantage of exciting new opportunities to network with colleagues, future employers, potential business partners, industry experts, and more than 50,000 ICC members. ICC also has over 300 chapters across North America and around the globe to help you stay informed on local events, to consult with other professionals, and to enhance your reputation in the local community.

JOIN NOW! 1-888-422-7233, x33804 | www.iccsafe.org/membership

People Helping People Build a Safer World™

09-01530